Algebra II Essentials FOR DUMMIES®

by Mary Jane Sterling

Wiley Publishing, Inc.

Algebra II Essentials For Dummies®

Published by
Wiley Publishing, Inc.
111 River St.
Hoboken, NJ 07030-5774
www.wiley.com

For general information on our other products and services, please contact our Customer Care Department within the U.S. at 877-762-2974, outside the U.S. at 317-572-3993, or fax 317-572-4002.

For technical support, please visit www.wiley.com/techsupport.

Wiley also publishes its books in a variety of electronic formats. Some content that appears in print may not be available in electronic books.

Library of Congress Control Number: 2010925242

ISBN: 978-0-470-61840-0

Manufactured in the United States of America

10 9 8 7 6 5 4 3 2 1

About the Author

Mary Jane Sterling has been an educator since graduating from college. Teaching at the junior high, high school, and college levels, she has had the full span of experiences and opportunities to determine how best to explain how mathematics works. She has been teaching at Bradley University in Peoria, Illinois, for the past 30 years. She is also the author of *Algebra II For Dummies, Trigonometry For Dummies, Math Word Problems For Dummies, Business Math For Dummies,* and *Linear Algebra For Dummies.*

Publisher's Acknowledgments

We're proud of this book; please send us your comments at `http://dummies.custhelp.com`. For other comments, please contact our Customer Care Department within the U.S. at 877-762-2974, outside the U.S. at 317-572-3993, or fax 317-572-4002.

Some of the people who helped bring this book to market include the following:

Acquisitions, Editorial, and Media Development

Project Editor: Elizabeth Kuball

Senior Acquisitions Editor: Lindsay Sandman Lefevere

Copy Editor: Elizabeth Kuball

Assistant Editor: Erin Calligan Mooney

Editorial Program Coordinator: Joe Niesen

Technical Editors: Tony Bedenikovic, Stefanie Long

Senior Editorial Manager: Jennifer Ehrlich

Editorial Supervisor and Reprint Editor: Carmen Krikorian

Senior Editorial Assistant: David Lutton

Editorial Assistants: Rachelle Amick, Jennette ElNaggar

Cover Photos: ©iStock

Cartoons: Rich Tennant (`www.the5thwave.com`)

Composition Services

Project Coordinator: Sheree Montgomery

Layout and Graphics: Carl Byers, Carrie A. Cesavice, Joyce Haughey, Mark Pinto

Proofreaders: Melissa Cossell, Tricia Liebig

Indexer: Potomac Indexing, LLC

Publishing and Editorial for Consumer Dummies

Diane Graves Steele, Vice President and Publisher, Consumer Dummies

Kristin Ferguson-Wagstaffe, Product Development Director, Consumer Dummies

Ensley Eikenburg, Associate Publisher, Travel

Kelly Regan, Editorial Director, Travel

Publishing for Technology Dummies

Andy Cummings, Vice President and Publisher, Dummies Technology/General User

Composition Services

Debbie Stailey, Director of Composition Services

Contents at a Glance

Table of Contents

Chapter 8: Being Respectful of Rational Functions . . . 91

Chapter 9: Examining Exponential and Logarithmic Functions . . . 107

Chapter 10: Getting Creative with Conics . . . 121

Introduction

Here you are, perusing a book on the essentials of Algebra II. You'll find here, as Joe Friday (star of the old *Dragnet* series) said, "The facts, ma'am, just the facts." For those of you too young to remember *Dragnet,* just think of this essentials book as being the Twitter version — not too detailed but with all the necessary information. In this book, you find the information you need with enough examples to show you the processes, but not a bunch of nitty-gritty details that tend to get in the way.

About This Book

A book on Algebra II isn't a romance novel (although I do love math), and it isn't science fiction. You could think of this book as a cross between a travel guide and a mathematical laboratory manual. How do travel and math go together? Let me try some situations that may fit:

- You just finished working through Algebra I and feel eager to embark on a new adventure.
- You haven't worked with algebra in a while, but math has always been your strength, so you think that a little prepping with some basic concepts will bring you up to speed.
- You're helping a friend or family member with Algebra II and want just the most necessary information — no frills or extra side-trips.

Even though I've pared the material in this book down to the basics, I haven't lost sight of the fact that other math areas are what drive Algebra II. Algebra is the passport to studying calculus, trigonometry, number theory, geometry, and all sorts of good mathematics. Algebra is basic, and the algebra you find here will help you grow your skills and knowledge so you can do well in math courses and possibly pursue other math topics.

Conventions Used in This Book

To help you navigate this book, I use the following conventions:

- I *italicize* special mathematical terms and define them right then and there so you don't have to search around.
- I use **boldface** text to indicate keywords in bulleted lists or the action parts of numbered steps. I describe many algebraic procedures in a step-by-step format and then use those steps in an example or two.

Foolish Assumptions

Algebra II is essentially a continuation of Algebra I, so I need to make some assumptions about readers of this book.

I assume that a person taking on Algebra II has a grasp of working with operations on signed numbers, simplifying radical expressions, and manipulating with rational terms. Another assumption I make is that your order of operations is in order. You should be able to work your way through algebraic equations and expressions using the ordering rules. I also assume that you know how to solve basic linear and quadratic equations and can make quick sketches of basic graphs. Even though I lightly cover these topics in this book, I assume that you have a general knowledge of the necessary procedures.

If you feel a bit over your head after reading through some chapters, you may want to refer to *Algebra I For Dummies,* 2nd Edition (Wiley), or *Algebra II For Dummies* (Wiley) for a more complete explanation of the basics. My feelings won't be hurt; I wrote those, too!

Icons Used in This Book

The icons that appear in this book are great for calling attention to the hot topics when doing algebra.

This icon provides you the rule or law or instruction on how to proceed whenever encountering the particular mathematical situation. The algebra rule given is "the law" — it always applies and always must be followed.

When you see the Example icon, you know that you'll find the result of an attempt at working out an equation or concept. An example often has a hidden agenda — it shows you more of a process than a basic rule can get across by itself.

This icon is like the sign alerting you to the presence of something special to watch out for on your adventure. It can save you time and energy. Use this information to cut to the chase and avoid unnecessary detours.

This icon helps you bring back information that you may have misplaced along the way. The information is needed to get you from here to the goal.

This icon alerts you to common hazards and stumbling blocks that could trip you up — cause accidents or get you into trouble with the math police. Those who have gone before you have found that these items can cause a big problem — so pay heed.

Where to Go from Here

You can use the table of contents at the beginning of the book and the index in the back to navigate your way to the topic that you're most interested in. You may want to start with some problem solving — in the form of equations or inequalities. If that's the case, then look at Chapter 2 for linear equations and inequalities or Chapters 3 and 4 for quadratic and other degree equations. Chapter 5 is your destination if you want to see what constitutes a function and its characteristics. And specific functions such as linear and quadratics are found in Chapter 6; polynomials are found in Chapter 7, rationals in Chapter 8, and exponentials and logs in Chapter 9. I saved the imaginary for last, in Chapter 12. But you could stop off and look at conics in Chapter 10, if those curves are of interest. Also, systems of equations incorporate several types of functions, and you find them in Chapter 11.

And, if you're more of a freewheeling type of guy or gal, take your finger, flip open the book, and mark a spot. No matter your motivation or what technique you use to jump into this book, you won't get lost because you can go in any direction from there.

Enjoy!

"We're both mathematicians, Sheldon, so let me explain it this way: Where r denotes the ordinate of our relationship at the time, t, above the point x, where b denotes boring, o denotes over, m denotes moving on. . . . Are you starting to get any of this?"

Chapter 1

Making Advances in Algebra

In This Chapter

- Making algebra orderly with the order of operations and other properties
- Enlisting rules of exponents
- Focusing on factoring

Algebra is a branch of mathematics that people study before they move on to other areas or branches in mathematics and science. Algebra all by itself is esthetically pleasing, but it springs to life when used in other applications.

Any study of science or mathematics involves rules and patterns. You approach the subject with the rules and patterns you already know, and you build on those rules with further study. In this chapter, I recap for you the basic rules from Algebra I so that you work from the correct structure. I present these basics so you can further your study of algebra and feel confident in your algebraic ability.

Bringing Out the Best in Algebraic Properties

Mathematicians developed the rules and properties you use in algebra so that every student, researcher, curious scholar, and bored geek working on the same problem would get the same answer — no matter the time or place.

Making short work of the basic properties

The commutative, associative, and other such properties are not only basic to algebra, but also to geometry and many other mathematical topics. I present the properties here so that I can refer to them as I solve equations and simplify expressions in later chapters.

The commutative property

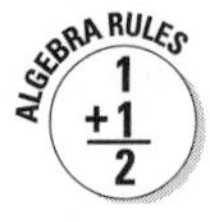

The *commutative property* applies to the operations of addition and multiplication. It states that you can change the order of the values in an operation without changing the final result:

$a + b = b + a$	Commutative property of addition
$a \cdot b = b \cdot a$	Commutative property of multiplication

So you can be sure that $2 + 4 = 4 + 2$ and $8 \cdot 7 = 7 \cdot 8$.

The associative property

Like the commutative property (see the preceding section), the *associative property* applies to the operations of addition and multiplication. The *associative property* states that you can change the grouping of operations without changing the result:

$a + (b + c) = (a + b) + c$	Associative property of addition
$a(b \cdot c) = (a \cdot b)c$	Associative property of multiplication

This property tells you that $3 + (8 + 5) = (3 + 8) + 5$ and that $-4 \cdot (8 \cdot 3) = (-4 \cdot 8) \cdot 3$.

The distributive property

The *distributive property* states that you can multiply each term in an expression within parentheses by the factor outside the parentheses and not change the value of the expression. It takes one operation — multiplication — and spreads it out over terms that you add to and subtract from one another:

$a(b + c - d) =$ $a \cdot b + a \cdot c - a \cdot d$	Distributing multiplication over addition and subtraction

For example, you can use the distributive property on the problem $12\left(\frac{1}{2}+\frac{2}{3}-\frac{3}{4}\right)$ to make your life easier. You distribute the 12 over the fractions by multiplying each fraction by 12 and then combining the results: $= 12\cdot\frac{1}{2}+12\cdot\frac{2}{3}-12\cdot\frac{3}{4}=6+8-9=5$.

Identities

The numbers 0 and 1 have special roles in algebra — as *identities.*

$a + 0 = 0 + a = a$	The *additive identity* is 0. Adding 0 to a number doesn't change that number; the number keeps its identity.
$a \cdot 1 = 1 \cdot a = a$	The *multiplicative identity* is 1. Multiplying a number by 1 doesn't change that number; the number keeps its identity.

Inverses

You find two types of *inverses* in algebra — additive inverses and multiplicative inverses:

- A number and its *additive inverse* add up to 0.
- A number and its *multiplicative inverse* have a product of 1.

The additive inverse of 6 is –6, so $6 + (-6) = 0$. And the multiplicative inverse of 6 is $\frac{1}{6}$, so $6\cdot\frac{1}{6}=1$.

The multiplication property of zero

The *multiplication property of zero* (MPZ) states that if the product of $a \cdot b \cdot c \cdot d \cdot e \cdot f = 0$, at least one of the terms has to represent the number 0. The only way the product of two or more values can be 0 is for at least one of the values to actually be 0. If you multiply (16)(467)(11)(9)(0), the result is 0. It doesn't really matter what the other numbers are — the 0 always wins.

Organizing your operations

When mathematicians switched from words to symbols to describe mathematical processes, their goal was to make dealing with problems as simple as possible; however, at the same time, they wanted everyone to know what was meant by an expression and for everyone to get the same answer to a problem. Along with the special notation came a special set of rules on how to handle more than one operation in an expression.

The *order of operations* dictates that you follow this sequence:

1. Raise to powers or find roots.
2. Multiply or divide.
3. Add or subtract.

If you have to perform more than one operation from the same level, work those operations moving from left to right. If any grouping symbols appear, perform the operation inside the grouping symbols first.

So, to do the problem $4+3^2-5\cdot 6+\sqrt{23-7}+\frac{14}{2}$, follow the order of operations:

1. The radical acts like a grouping symbol, so you subtract what's in the radical first to get $4+3^2-5\cdot 6+\sqrt{16}+\frac{14}{2}$.
2. Raise the power and find the root: $4+9-5\cdot 6+4+\frac{14}{2}$.
3. Do the multiplication and then the division: $4+9-30+4+7$.
4. Add and subtract, moving from left to right: $4+9-30+4+7=13-30+4+7=-17+4+7=-13+7=-6$.

Enumerating Exponential Rules

Several hundred years ago, mathematicians introduced powers of variables and numbers called *exponents.* Instead of writing *xxxxxxxx*, you use the exponent 8 by writing x^8. This form is easier to read and much quicker. The use of exponents expanded to being able to write fractions with negative exponents and radicals with fractional exponents. You find all the details in *Algebra I For Dummies,* 2nd Edition (Wiley).

Multiplying and dividing exponents

When two numbers or variables have the same base, you can multiply or divide those numbers or variables by adding or subtracting their exponents:

- $a^m \cdot a^n = a^{m+n}$: When multiplying numbers with the same base, you add the exponents.
- $\frac{a^m}{a^n} = a^{m-n}$: When dividing numbers with the same base, you subtract the exponents (numerator minus denominator).

To multiply $x^4 \cdot x^5$, for example, you add: $x^{4+5} = x^9$. When dividing x^8 by x^5, you subtract: $\frac{x^8}{x^5} = x^{8-5} = x^3$.

You have to be sure that the bases of the expressions are the same. You can multiply or divide 3^2 and 3^4, but you can't use the multiplication or division rules for exponents to multiply or divide 3^2 and 4^3.

Rooting out exponents

Radical expressions — such as square roots, cube roots, fourth roots, and so on — appear with a radical to show the root. Another way you can write these values is by using fractional exponents. You'll have an easier time combining variables with the same base if they have fractional exponents in place of radical forms:

- $\sqrt[n]{x} = x^{\frac{1}{n}}$: The root goes in the denominator of the fractional exponent.
- $\sqrt[n]{x^m} = x^{\frac{m}{n}}$: The root goes in the denominator of the fractional exponent, and the power goes in the numerator.

To simplify a radical expression such as $\frac{\sqrt[4]{x}\sqrt[6]{x^5}}{\sqrt[3]{x}}$, you change the radicals to exponents and apply the rules for multiplication and division of values with the same base (see the preceding section):

$$\begin{aligned}\frac{\sqrt[4]{x}\sqrt[6]{x^5}}{\sqrt[3]{x}} &= \frac{x^{\frac{1}{4}} \cdot x^{\frac{5}{6}}}{x^{\frac{1}{3}}} \\ &= x^{\frac{1}{4}+\frac{5}{6}-\frac{1}{3}} \\ &= x^{\frac{3}{12}+\frac{10}{12}-\frac{4}{12}} \\ &= x^{\frac{9}{12}} \\ &= x^{\frac{3}{4}} \\ &= \sqrt[4]{x^3}\end{aligned}$$

Powering up exponents

When raising a power to a power, you multiply the exponents. When taking the root of a power, you divide the exponent by the root:

- $(a^m)^n = a^{(m)(n)}$: Raise a power to a power by multiplying the exponents.
- $\sqrt[n]{a^m} = a^{\frac{m}{n}}$: Reduce the power when taking a root by dividing the exponents.

The second rule may look familiar — it's one of the rules that govern changing from radicals to fractional exponents from the preceding section. Here's an example of how you apply the two rules when simplifying an expression:

$$\sqrt[3]{\left(x^4\right)^6 \cdot x^9} = \sqrt[3]{x^{24} \cdot x^9} = \sqrt[3]{x^{33}} = x^{\frac{33}{3}} = x^{11}$$

Working with negative exponents

You use negative exponents to indicate that a number or variable belongs in the denominator of the term:

$$\frac{1}{a} = a^{-1} \qquad \frac{1}{a^n} = a^{-n} \qquad \frac{1}{a^{-n}} = a^{-(-n)} = a^n$$

Writing variables with negative exponents allows you to combine those variables with other factors that share the same base. You can rewrite the fractions by using negative exponents and then simplify by using the rules for multiplying factors with the same base:

$$\frac{1}{x^4}\cdot x^7\cdot\frac{3}{x} = x^{-4}\cdot x^7\cdot 3x^{-1} = 3x^{-4+7-1} = 3x^2$$

Assigning Factoring Techniques

When you *factor* an algebraic expression, you rewrite the sums and differences of the terms as a product. The factored form comes in handy when you set an expression equal to 0 to solve an equation. Factored numerators and denominators in fractions also make it possible to reduce the fractions.

Making two terms factor

When an algebraic expression has two terms, you have four different choices for its factorization — if you can factor the expression at all. If you try the following four methods and none of them works, you can stop your attempt; you just can't factor the expression:

$ax + ay = a(x + y)$	Greatest common factor (GCF)
$x^2 - a^2 = (x - a)(x + a)$	Difference of two perfect squares
$x^3 - a^3 = (x - a)(x^2 + ax + a^2)$	Difference of two perfect cubes
$x^3 + a^3 = (x + a)(x^2 - ax + a^2)$	Sum of two perfect cubes

In general, you check for a GCF before attempting any of the other methods. By taking out the common factor, you often make the numbers smaller and more manageable, which helps you see clearly whether any other factoring is necessary.

Factor the expression $6x^4 - 6x$.

First factor out the common factor, $6x$, and then use the pattern for the difference of two perfect cubes:

$$6x^4 - 6x = 6x(x^3 - 1) = 6x(x - 1)(x^2 + x + 1)$$

A *quadratic trinomial* is a three-term polynomial with a term raised to the second power (and no higher powers). When you see something like $x^2 + x + 1$ (as in this case), you immediately run through the possibilities of factoring it into the product of two binomials (see the next section). You can just stop. These trinomials that crop up with factoring cubes just don't cooperate.

Factor $48x^3y^2 - 300x^3$.

When you factor the expression, first divide out the common factor, $12x^3$, to get $12x^3(4y^2 - 25)$. Then factor the difference of perfect squares in the parentheses: $48x^3y^2 - 300x^3 = 12x^3(2y - 5)(2y + 5)$.

Factoring three terms

When a quadratic expression has three terms, making it a *trinomial*, you have two different ways to factor it. One method is factoring out a GCF, and the other is finding two binomials whose product is identical to the sum and/or difference of the original three terms:

$$ax + ay + az = a(x + y + z) \qquad \text{GCF}$$

$$ax^{2n} + bx^n + c = (dx^n + e)(fx^n + g) \qquad \text{Two binomials}$$

When you factor a trinomial that results from multiplying two binomials, you have to play detective and piece together the parts of the puzzle. Look at the following generalized product of binomials and the pattern that appears:

$$(dx + e)(fx + g) = dfx^2 + dgx + efx + eg = dfx^2 + (dg + ef)x + eg$$

$$= ax^2 + bx + c$$

The F in FOIL stands for *first;* the *first* terms are the dx and fx. The O in FOIL stands for *outer;* the *outer* terms are dx and g. The I in FOIL stands for *inner;* the *inner* terms are e and fx.

Their products are dgx (outer) and efx (inner). You add these two values. The L in FOIL stands for *last;* the *last* terms, e and g, have a product of eg.

Now, think of every quadratic trinomial as being of the form $ax^2 + bx + c = dfx^2 + (dg + ef)x + eg$. The coefficient of the x^2 term, df, is the product of the coefficients of the two x terms in the parentheses; the last term, eg, is the product of the two second terms in the parentheses; and the coefficient of the middle term is the sum of the outer and inner products. To factor these trinomials into the product of two binomials, you use the opposite of FOIL and figure out which factorizations to use.

Here are the basic steps you take to unFOIL a quadratic trinomial:

1. **Determine all the ways you can multiply two numbers to get a, the coefficient of the squared term.**
2. **Determine all the ways you can multiply two numbers to get c, the constant term.**
3. **If the last term is positive, find the combination of factors from steps 1 and 2 whose *sum* is that middle term; if the last term is negative, you want the combination to be a difference.**
4. **Arrange your choices as binomials so that the factors line up correctly.**
5. **Insert the + and – signs to finish off the factoring and make the sign of the middle term come out right.**

Factor $x^2 + 9x + 20$.

You find two terms whose product is 20 and whose sum is 9. The coefficient of the squared term is 1, so you don't have to take any other factors into consideration. You choose 4 and 5 as the factors of 20, because $4 + 5 = 9$. Arranging the factors and x's, you get $x^2 + 9x + 20 = (x + 4)(x + 5)$.

Factor $6x^2 - x - 12$.

You have to consider both the factors of 6 and the factors of 12. Start 2 and 3 for the factors of 6 and write: $(2x \quad)(3x \quad)$. Don't insert any signs until the end of the process.

Now, using the factors of 12, you look for a pairing that gives you a difference of 1 between the outer and inner products. Try the product of $3 \cdot 4$, matching (multiplying) the 3 with the $3x$ and the 4 with the $2x$. Bingo! Write $(2x \quad 3)(3x \quad 4)$. You'll multiply the 3 and $3x$ because they're in different parentheses — not the same one. The difference has to be negative, so you can put the negative sign in front of the 3 in the first binomial: $6x^2 - x - 12 = (2x - 3)(3x + 4)$.

Factoring four or more terms by grouping

When four or more terms come together to form an expression, you look for a GCF first. If you can't find a factor common to all the terms at the same time, your other option is *grouping*. To group, you take the terms two at a time and look for common factors for each of the pairs on an individual basis. After factoring, you see if the new groupings have a common factor.

Factor $x^3 - 4x^2 + 3x - 12$.

The four terms of $x^3 - 4x^2 + 3x - 12$ don't have any common factor. However, the first two terms have a common factor of x^2, and the last two terms have a common factor of 3:

$$x^3 - 4x^2 + 3x - 12 = x^2(x - 4) + 3(x - 4)$$

Notice that you now have two terms, not four, and they both have the factor $(x - 4)$. Now, factoring $(x - 4)$ out of each term, you have $(x - 4)(x^2 + 3)$.

Chapter 2

Lining Up Linear Equations

In This Chapter

- Establishing a game plan for solving linear equations
- Working through special rules for linear inequalities
- Making short work of absolute value equations and inequalities

The term *linear* has the word *line* buried in it, and the obvious connection is that you can graph many linear equations as lines. But linear expressions can come in many types of packages, not just equations or lines. In this chapter, you find out how to deal with linear equations, what to do with the answers in linear inequalities, and how to rewrite linear absolute value equations and inequalities so that you can solve them.

Getting the First Degree: Linear Equations

Linear equations feature variables that reach only the first degree, meaning that the highest power of any variable you solve for is 1. The general form of a linear equation with one variable is $ax + b = c$.

The one variable is the x. But, no matter how many variables you see, the common theme to linear equations is that each variable has only one solution or value that satisfies the equation when matched with constants or specific other variables.

Solving basic linear equations

To solve a linear equation, you isolate the variable on one side of the equation by adding the same number to both sides — or you can subtract, multiply, or divide the same number on both sides.

For example, you solve the equation $4x - 7 = 21$ by adding 7 to each side of the equation, to isolate the variable and the multiplier, and then dividing each side by 4, to leave the variable on its own:

$$4x - 7 + 7 = 21 + 7 \quad \rightarrow \quad 4x = 28$$

$$4x \div 4 = 28 \div 4 \quad \rightarrow \quad x = 7$$

When a linear equation has grouping symbols such as parentheses, brackets, or braces, you deal with any distributing across and simplifying within the grouping symbols before you isolate the variable. For example, to solve the equation $5x - [3(x + 2) - 4(5 - 2x) + 6] = 20$, you first distribute the 3 and –4 inside the brackets:

$$5x - [3x + 6 - 20 + 8x + 6] = 20$$

Then you combine the terms that combine and distribute the negative sign (–) in front of the bracket; it's like multiplying through by –1:

$$\begin{aligned} 5x - [11x - 8] &= 20 \\ 5x - 11x + 8 &= 20 \end{aligned}$$

Simplify again, and you can solve for x:

$$\begin{aligned} -6x + 8 &= 20 \\ -6x &= 12 \\ x &= -2 \end{aligned}$$

Eliminating fractions

The problem with fractions, like cats, is that they aren't particularly easy to deal with. They always insist on having their own way — in the form of common denominators before you can add or subtract. And division? Don't get me started!

The best way to deal with linear equations that involve variables tangled up with fractions is to get rid of the fractions. Your game plan is to multiply both sides of the equation by the least common denominator of all the fractions in the equation.

Solve $\frac{x+2}{5}+\frac{4x+2}{7}=\frac{9-x}{2}$ for x.

Multiply each term in the equation by 70 — the least common denominator (also known as the *least common multiple*) for fractions with the denominators 5, 7, and 2:

$$^{14}\cancel{70}\left(\frac{x+2}{\cancel{5}_1}\right)+{}^{10}\cancel{70}\left(\frac{4x+2}{\cancel{7}_1}\right)={}^{35}\cancel{70}\left(\frac{9-x}{\cancel{2}_1}\right)$$

Now you distribute the reduced numbers over each set of parentheses, combine the like terms, and solve for x:

$$\begin{aligned}14x+28+40x+20&=315-35x\\54x+48&=315-35x\\89x&=267\\x&=3\end{aligned}$$

Lining Up Linear Inequalities

Algebraic *inequalities* show comparative relationships between a number and an expression or between two expressions. In other words, you use inequalities for comparisons.

Inequalities in algebra are expressed by the comparisons *less than* ($<$), *greater than* ($>$), *less than or equal to* ($\leq$), and *greater than or equal to* ($\geq$). A linear equation containing one variable has only one solution, but a linear inequality can have an infinite number of solutions.

Here are the rules for operating on inequalities (you can replace the $<$ symbol with any of the inequality symbols, and the rules will still hold):

- If $a < b$, then $a + c < b + c$ (adding any number).
- If $a < b$, then $a - c < b - c$ (subtracting any number).
- If $a < b$ and $c > 0$, then $a \cdot c < b \cdot c$ (multiplying by any *positive* number).

- If $a < b$ and $c < 0$, then $a \cdot c > b \cdot c$ (multiplying by any *negative* number).
- If $a < b$ and $c > 0$, then $\frac{a}{c} < \frac{b}{c}$ (dividing by any *positive* number).
- If $a < b$ and $c < 0$, then $\frac{a}{c} > \frac{b}{c}$ (dividing by any *negative* number).
- If $\frac{a}{c} < \frac{b}{d}$, then $\frac{c}{a} > \frac{d}{b}$ (reciprocating fractions).

You must not multiply or divide an inequality by 0.

Solving basic inequalities

To solve a basic linear inequality, first move all the variable terms to one side of the inequality and the numbers to the other. After you simplify the inequality down to a variable and a number, you can find out what values of the variable will make the inequality into a true statement.

Solve $3x + 4 > 11 - 4x$ for x.

Add $4x$ and subtract 4 from each side: $7x > 7$.

Divide each side by 7: $x > 1$.

The sense stayed the same, because you didn't multiply or divide each side by a negative number.

The rules for solving linear equations also work with inequalities — somewhat. Everything goes smoothly until you try to multiply or divide each side of an inequality by a negative number.

When you multiply or divide each side of an inequality by a negative number, you have to *reverse the sense* (change $<$ to $>$, or vice versa) to keep the inequality true.

Solve the inequality $4(x - 3) - 2 \geq 3(2x + 1) + 7$ for x.

Distributing, you get: $4x - 12 - 2 \geq 6x + 3 + 7$.

Simplifying: $4x - 14 \geq 6x + 10$.

Now subtract $6x$ and add 14: $-2x \geq 24$.

Divide each side by –2, reversing the sense: $x \leq -12$.

Introducing interval notation

Much of higher mathematics uses interval notation instead of inequality notation. Interval notation is thought to be quicker and neater than inequality notation. Interval notation uses parentheses, brackets, commas, and the infinity symbol to bring clarity to the murky inequality waters.

To use *interval notation* when describing a set of numbers:

- You order any numbers used in the notation with the smaller number to the left of the larger number.
- You indicate "or equal to" by using a bracket.
- If the solution doesn't include the end number, you use a parenthesis.
- When the interval doesn't end (it goes up to positive infinity or down to negative infinity), use $+\infty$ or $-\infty$, whichever is appropriate, and a parenthesis.

Here are some examples of inequality notation and the corresponding interval notation:

Inequality Notation	*Linear Notation*
$x < 3$	$(-\infty, 3)$
$x \geq -2$	$[-2, \infty)$
$4 \leq x < 9$	$[4, 9)$
$-3 < x < 7$	$(-3, 7)$

Solve the inequality $-8 \le 3x - 5 < 10$.

Add 5 to each of the three sections and then divide each section by 3:

$$\begin{array}{l}
-8 \le 3x - 5 < 10 \\
\underline{+5 \qquad +5 \quad +5} \\
-3 \le 3x \quad < 15 \\
\frac{-3}{3} \le \frac{3x}{3} \quad < \frac{15}{3} \\
-1 \le x \quad < 5
\end{array}$$

You write the answer, $-1 \le x < 5$, in interval notation as $[-1, 5)$.

Absolute Value: Keeping Everything in Line

When you perform an *absolute value operation,* you're not performing surgery at bargain-basement prices; you're taking a number inserted between the absolute value bars, $|a|$, and recording the distance of that number from 0 on the number line. For example, $|3| = 3$, because 3 is three units away from 0. On the other hand, $|-4| = 4$, because –4 is four units away from 0.

The absolute value of a is defined as

$$|a| = \begin{cases} a \text{ if } a \ge 0 \\ -a \text{ if } a < 0 \end{cases}$$

You read the definition as follows: "The absolute value of a is equal to a, itself, if a is positive or 0; the absolute value of a is equal to the *opposite* of a if a is negative."

Solving absolute value equations

A linear absolute value equation is an equation that takes the form $|ax + b| = c$. To solve an absolute value equation in this linear form, you have to consider both possibilities: $ax + b$ may be positive, or it may be negative.

To solve for the variable x in $|ax + b| = c$, you solve both $ax + b = c$ and $ax + b = -c$.

Solve the absolute value equation $3|4 - 3x| + 7 = 25$.

First, you have to subtract 7 from each side of the equation and then divide each side by 3:

$$3|4 - 3x| + 7 = 25$$
$$3|4 - 3x| = 18$$
$$|4 - 3x| = 6$$

Then you can apply the rule for changing the absolute value equation to two linear equations:

$$4 - 3x = 6 \qquad\qquad 4 - 3x = -6$$
$$-3x = 2 \qquad\qquad -3x = -10$$
$$x = -\frac{2}{3} \qquad\qquad x = \frac{10}{3}$$

Seeing through absolute value inequality

An absolute value inequality contains an absolute value, $|a|$, and an inequality: $<$, $>$, $\leq$, or $\geq$.

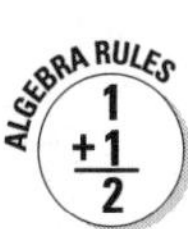

To solve an absolute value inequality, you have to change the form from absolute value to just plain inequality.

- To solve for x in $|ax + b| < c$, you solve $-c < ax + b < c$.
- To solve for x in $|ax + b| > c$, you solve $ax + b > c$ and $ax + b < -c$.

Solve the absolute value inequality: $|2x - 1| \leq 5$.

Rewrite the inequality: $-5 \leq 2x - 1 \leq 5$.

Next, add 1 to each section: $-4 \leq 2x \leq 6$.

Divide each section by 2: $-2 \leq x \leq 3$.

You can write the solution in interval notation as $[-2, 3]$.

Solve $|7-2x|>11$ for x.

Rewrite the absolute value inequality as two separate inequalities: $7-2x>11$ and $7-2x<-11$.

When solving the two inequalities, be sure to remember to switch the sign when you divide by –2:

$$\begin{aligned} 7-2x&>11 \\ -2x&>4 \\ x&<-2 \end{aligned} \qquad\qquad \begin{aligned} 7-2x&<-11 \\ -2x&<-18 \\ x&>9 \end{aligned}$$

The solution $x<-2$ or $x>9$, in interval notation, is $(-\infty,-2)$ or $(9,\infty)$.

Don't write the solution $x<-2$ or $x>9$ as $9<x<-2$. If you do, you indicate that some numbers can be bigger than 9 *and* smaller than –2 at the same time, which just isn't so.

Chapter 3

Making Quick Work of Quadratic Equations

In This Chapter

- Solving quadratic equations by factoring or taking roots
- Using the quadratic formula
- Coming to grips with quadratic inequalities

A *quadratic equation* contains a variable term with an exponent of 2, and no term with a higher power. The standard form is $ax^2 + bx + c = 0$. Quadratic equations potentially have two real solutions. You may not find two, but you start out assuming that you'll find two and then proceed to prove or disprove your assumption. Quadratic equations also serve as good models for practical applications.

In this chapter, you discover many ways to approach both simple and advanced quadratic equations. You can solve some quadratic equations in only one way, and you can solve others by readers' choice (factoring, quadratic formula, or by-guess-or-by-golly) — whatever your preference. It's nice to be able to choose. But if you have a choice, I hope you choose the quickest and easiest ways possible, so I cover these first in this chapter (except for by-guess-or-by-golly).

Using the Square Root Rule When Possible

Some quadratic equations are easier to solve than others. Half the battle is recognizing which equations are easy and which are more challenging.

If a quadratic equation is made up of a squared term and a number, written in the form $x^2 = k$, you solve the equation using the *square root rule:* If $x^2 = k$, $x = \pm\sqrt{k}$.

The number represented by k has to be positive if you want to find real answers with this rule. If k is negative, you get an imaginary answer, such as $3i$ or $5 - 4i$. (For more on imaginary numbers, check out Chapter 12.)

Solve for x using the square root rule: $6x^2 = 96$.

The initial equation doesn't strictly follow the format for the square root rule because of the coefficient 6, but you can get to the proper form pretty quickly. You divide each side of the equation by the coefficient; in this case, you get $x^2 = 16$; now you're in business. Taking the square root of each side, you get $x = \pm 4$.

Solve $y^2 = 40$ for y.

$$\begin{aligned} y^2 &= 40 \\ y &= \pm\sqrt{40} \\ &= \pm\sqrt{4}\sqrt{10} \\ &= \pm 2\sqrt{10} \end{aligned}$$

You can use a law of radicals to simplify a radical expression. Separate the number under the radical into two factors — one of which is a perfect square: $\sqrt{a \cdot b} = \sqrt{a}\sqrt{b}$.

Solving Quadratic Equations by Factoring

When you can *factor* a quadratic expression that's part of a quadratic equation you can solve quadratic equations by setting the factored expression equal to zero (making it an equation) and then using the multiplication property of zero (MPZ; see Chapter 1). How you factor the expression depends on the number of terms in the quadratic and how those terms are related.

Factoring quadratic binomials

You can factor a quadratic binomial in one of two ways — if you can factor it at all (you may find no common factor, or the two terms may not both be squares):

- Divide out a common factor from each of the terms.
- Write the quadratic as the product of two binomials, if the quadratic is the difference of perfect squares.

Taking out a greatest common factor

The *greatest common factor* (GCF) of two or more terms is the largest number (and variable combination) that divides each of the terms evenly.

Solve the equation $4x^2 + 8x = 0$ using factorization and the MPZ.

Factor out the GCF: $4x(x + 2) = 0$.

Using the MPZ, you can now make one of three statements about this equation:

- $4 = 0$, which is false — this isn't a solution
- $x = 0$
- $x + 2 = 0$, which means that $x = -2$

You find two solutions for the original equation $4x^2 + 8x = 0$: $x = 0$ or $x = -2$. If you replace the x's with either of these solutions, you create a true statement.

Be careful when the GCF of an expression is just x, and always remember to set that front factor, x, equal to 0 so you don't lose one of your solutions. A really common error in algebra is to take a perfectly nice equation such as $x^2 + 5x = 0$, factor it into $x(x + 5) = 0$, and give the answer $x = -5$. Don't forget the solution $x = 0$!

Factoring the difference of squares

Use the factorization of the difference of squares to solve some quadratic equations.

This method states that if $x^2 - a^2 = 0$, $(x - a)(x + a) = 0$, and $x = a$ or $x = -a$. Generally, if $k^2x^2 - a^2 = 0$, $(kx - a)(kx + a) = 0$, and $x = \frac{a}{k}$ or $x = -\frac{a}{k}$.

Solve $49y^2 - 64 = 0$ using factorization and the MPZ.

Factor the terms on the left: $(7y - 8)(7y + 8) = 0$.

And using the MPZ, $y = \frac{8}{7}$ or $y = -\frac{8}{7}$.

Factoring quadratic trinomials

Like quadratic binomials, a quadratic trinomial can have as many as two solutions — or it may have one solution or no solution at all. If you can factor the trinomial and use the MPZ to solve for the roots, you're home free. If the trinomial doesn't factor, or if you can't figure out how to factor it, you can utilize the *quadratic formula* (see the section "The Quadratic Formula to the Rescue," later in this chapter). The rest of this section deals with the trinomials that you *can* factor.

Solve $x^2 - 2x - 15 = 0$ for x.

You can factor the left side of the equation into $(x - 5)(x + 3) = 0$ and then set each factor equal to 0. When $x - 5 = 0$, $x = 5$, and when $x + 3 = 0$, $x = -3$.

Solve $24x^2 + 52x - 112 = 0$ for x.

It may not be immediately apparent how you should factor such a seemingly complicated trinomial. But factoring 4 out of each term to simplify the picture a bit, you get $4(6x^2 + 13x - 28) = 0$.

Factoring the quadratic in the parentheses: $4(3x - 4)(2x + 7) = 0$. Setting the two binomials equal to 0, you get $x = \frac{4}{3}$ or $x = -\frac{7}{2}$.

The Quadratic Formula to the Rescue

The quadratic formula is a wonderful tool to use when other factoring methods fail (see the previous section). You take the numbers from a quadratic equation, plug them into the formula, and out come the solutions of the equation. You can even use the formula when the equation does factor, but you don't see how.

The *quadratic formula* states that when you have a quadratic equation in the form $ax^2 + bx + c = 0$ (where $a \neq 0$), the equation has the solutions $x = \frac{-b \pm \sqrt{b^2 - 4ac}}{2a}$.

Realizing rational solutions

You can factor quadratic equations such as $48x^2 - 155x + 125 = 0$ to find their solutions, but the factorization may not leap right out at you when the numbers are so large. Using the quadratic formula for this example, you let $a = 48$, $b = -155$, and $c = 125$. Filling in the values and solving for x, you get

$$\begin{aligned} x &= \frac{-(-155) \pm \sqrt{(-155)^2 - 4(48)(125)}}{2(48)} \\ &= \frac{155 \pm \sqrt{24{,}025 - 24{,}000}}{96} \\ &= \frac{155 \pm \sqrt{25}}{96} \\ &= \frac{155 \pm 5}{96} \end{aligned}$$

Starting with the plus sign, the first solution is $\frac{155+5}{96} = \frac{160}{96} = \frac{5}{3}$. For the minus sign, you get $\frac{155-5}{96} = \frac{150}{96} = \frac{25}{16}$. The fact that you get fractions tells you that you could've factored the quadratic: $48x^2 - 155x + 125 = (3x - 5)(16x - 25) = 0$. Do you see where the 3 and 5 and the 16 and 25 come from in the answers?

Investigating irrational solutions

The quadratic formula is especially valuable for solving quadratic equations that don't factor. Unfactorable equations, when they do have real solutions, have irrational numbers in their answers. *Irrational numbers* have no fractional equivalent; they feature decimal values that go on forever and never have patterns that repeat.

Solve the quadratic equation $2x^2 + 5x - 6 = 0$.

Using the quadratic formula, you let $a = 2$, $b = 5$, and $c = -6$, to get the following:

$$\begin{aligned} x &= \frac{-5 \pm \sqrt{5^2 - 4(2)(-6)}}{2(2)} \\ &= \frac{-5 \pm \sqrt{25 + 48}}{4} \\ &= \frac{-5 \pm \sqrt{73}}{4} \end{aligned}$$

The answer $\frac{-5 + \sqrt{73}}{4}$ is approximately 0.886, and $\frac{-5 - \sqrt{73}}{4}$ is approximately –3.386. You find perfectly good answers, rounded off to the nearest thousandth. The fact that the number under the radical isn't a perfect square tells you something else: You couldn't have factored the quadratic, no matter how hard you tried.

Promoting Quadratic-like Equations

A *quadratic-like trinomial* is a trinomial of the form $ax^{2n} + bx^n + c = 0$. The power on one variable term is twice that of the other variable term, and a constant term completes the picture. The good thing about quadratic-like trinomials is that they're candidates for factoring and then for the application of the MPZ. Solve these equations by factoring the trinomials into the product of binomials and then applying the MPZ.

Solve: $z^6 - 26z^3 - 27 = 0$.

You can think of this equation as being like the quadratic $x^2 - 26x - 27$, which factors into $(x - 27)(x + 1)$. If you replace the x's in the factorization with z^3, you have the factorization for the equation with the z's:

$$z^6 - 26z^3 - 27 = (z^3 - 27)(z^3 + 1) = 0$$

Setting each factor equal to 0:

$$\begin{aligned} z^3 - 27 &= 0 \qquad & z^3 + 1 &= 0 \\ z^3 &= 27 & z^3 &= -1 \\ z &= 3 & z &= -1 \end{aligned}$$

Solve the quadratic-like trinomial $y^4 - 17y^2 + 16 = 0$.

Factor the trinomial into the product of two binomials. Then factor each binomial using the rule for the difference of squares:

$$\begin{aligned} y^4 - 17y^2 + 16 &= (y^2 - 16)(y^2 - 1) \\ &= (y - 4)(y + 4)(y - 1)(y + 1) \end{aligned}$$

Setting the individual factors equal to 0, you get $y = 4$, $y = -4$, $y = 1$, $y = -1$.

Solving Quadratic Inequalities

A *quadratic inequality* is just what it says: an inequality (<, >, ≤, or ≥) that involves a quadratic expression. You can employ the same method you use to solve a quadratic inequality to solve high-degree inequalities and rational inequalities (which contain variables in fractions).

You need to be able to solve quadratic equations in order to solve quadratic inequalities. With quadratic equations, you set the expressions equal to 0; inequalities use the same numbers that give you zeros and then determine what's on either side of the numbers (positives and negatives).

To solve a quadratic inequality, follow these steps:

1. **Move all the terms to one side of the inequality sign.**
2. **Factor, if possible.**
3. **Determine all zeros (roots or solutions).**

   Zeros are the values of x that make each factored expression equal to 0.
4. **Put the zeros in order on a number line.**
5. **Create a sign line to show where the expression in the inequality is positive or negative.**

   A *sign line* shows the signs of the different factors in each interval. If the expression is factored, show the signs of the individual factors.
6. **Determine the solution, writing it in inequality notation or interval notation (see Chapter 2).**

Keeping it strictly quadratic

The techniques you use to solve the inequalities in this section are also applicable for solving higher-degree polynomial inequalities and rational inequalities. If you can factor a third- or fourth-degree polynomial (see "Promoting Quadratic-like Equations" to get started), you can handily solve an inequality where the polynomial is set less than 0 or greater than 0. You can also use the sign-line method to look at factors of rational (fractional) expressions. For now, however, consider sticking to the quadratic inequalities.

To solve the inequality $x^2 - x > 12$, for example, you need to determine what values of x you can square so that when you subtract the original number, your answer will be bigger than 12. For example, when $x = 5$, you get $25 - 5 = 20$. That's certainly bigger than 12, so the number 5 works; $x = 5$ is a solution. How about the number 2? When $x = 2$, you get $4 - 2 = 2$, which isn't bigger than 12. You can't use $x = 2$ in the solution. Do you then conclude that smaller numbers don't work? Not so. When you try $x = -10$, you get $100 + 10 = 110$, which is most definitely bigger than 12. You can actually find infinitely many numbers that make this inequality a true statement.

Therefore, you need to solve the inequality by using the steps I outline in the introduction to this section:

1. **Subtract 12 from each side of the inequality $x^2 - x > 12$ to move all the terms to one side.**

   You end up with $x^2 - x - 12 > 0$.

2. **Factoring on the left side of the inequality, you get $(x - 4)(x + 3) > 0$.**

3. **Determine that all the zeroes for the inequality are $x = 4$ and $x = -3$.**

4. **Put the zeros in order on a number line, shown in the following figure.**

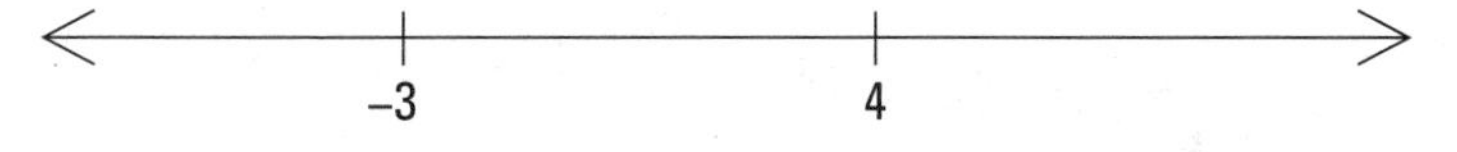

5. **Create a sign line to show the signs of the different factors in each interval.**

   Between –3 and 4, try letting $x = 0$ (you can use any number between –3 and 4). When $x = 0$, the factor $(x - 4)$ is negative, and the factor $(x + 3)$ is positive. Put those signs on the sign line to correspond to the factors. Do the same for the interval of numbers to the left of –3 and to the right of 4 (see the following figure).

   $(x-4)(x+3)$ $x = -5$ $(-)(-)$ | $(x-4)(x+3)$ $x = 0$ $(-)(+)$ | $(x-4)(x+3)$ $x = 10$ $(+)(+)$

   –3 4

   The x values in each interval are really random choices (as you can see from my choice of $x = -5$ and $x = 10$). Any number in each of the intervals gives you the same positive or negative value to the factor.

6. **To determine the solution, look at the signs of the factors; you want the expression to be positive, corresponding to the inequality *greater than zero.***

The interval to the left of –3 has a negative times a negative, which is positive. So, any number to the left of –3 works. You can write that part of the solution as $x < -3$ or, in interval notation, $(-\infty, -3)$. The interval to the right of 4 has a positive times a positive, which is positive. So, $x > 4$ is a solution; you can write it as $(4, \infty)$. No matter what numbers you choose in the interval between –3 and 4, the result is always negative because you have a negative times a positive. The complete solution lists both intervals that have working values in the inequality.

The solution of the inequality $x^2 - x > 12$, therefore, is $x < -3$ or $x > 4$.

Signing up for fractions

The sign-line process (see the introduction to this section and the previous example problem) is great for solving rational inequalities, such as $\frac{x-2}{x+6} \le 0$. The signs of the results of multiplication and division use the same rules, so to determine your answer, you can treat the numerator and denominator the same way you treat two different factors in multiplication.

Using the steps from the list I present in the introduction to this section, determine the solution for a rational inequality:

1. **Every term in $\frac{x-2}{x+6} \le 0$ is to the left of the inequality sign.**
2. **Neither the numerator nor the denominator factors any further.**
3. **The two zeros are $x = 2$ and $x = -6$.**
4. **You can see the two numbers on a number line in the following illustration.**

–6 2

5. **Create a sign line for the two zeroes.**

   You can see in the following figure that the numerator is positive when x is greater than 2, and the denominator is positive when x is greater than –6.

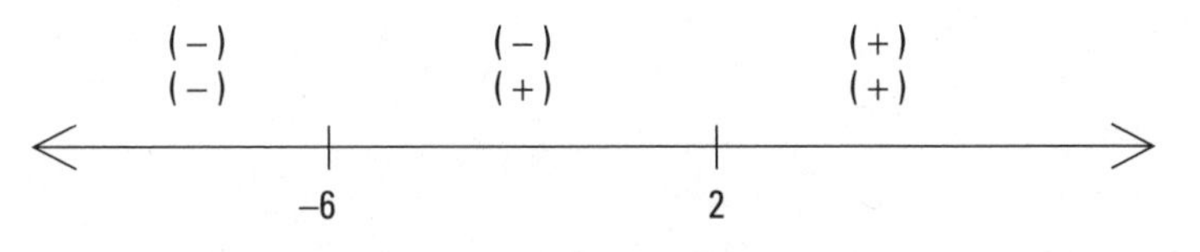

6. **When determining the solution, keep in mind that the inequality calls for something less than or equal to zero.**

   The fraction is a negative number when you choose an x between –6 and 2. You get a negative numerator and a positive denominator, which gives a negative result. Another solution to the original inequality is the number 2. Letting $x = 2$, you get a numerator equal to 0, which you want because the inequality is less than or equal to 0. You can't let the denominator be 0, though. Having a zero in the denominator isn't allowed because no such number exists. So, the solution of $\frac{x-2}{x+6} \le 0$ is $-6 < x \le 2$. In interval notation, you write the solution as $(-6, 2]$.

Increasing the number of factors

The method you use to solve a quadratic inequality (see the "Keeping it strictly quadratic" section, earlier in this chapter) works nicely with fractions and high-degree expressions. For example, you can solve $(x + 2)(x - 4)(x + 7)(x - 5)^2 \ge 0$ by creating a sign line and checking the products.

The inequality is already factored, so you move to the step (Step 3) where you determine the zeros. The zeros are –2, 4, –7, and 5 (the 5 is a double root and the factor is always positive or 0). The following figure shows the values in order on the number line.

Now you choose a number in each interval, substitute the numbers into the expression on the left of the inequality, and determine the signs of the four factors in those intervals. You can see from the following figure that the last factor, $(x-5)^2$, is always positive or 0, so that's an easy factor to pinpoint.

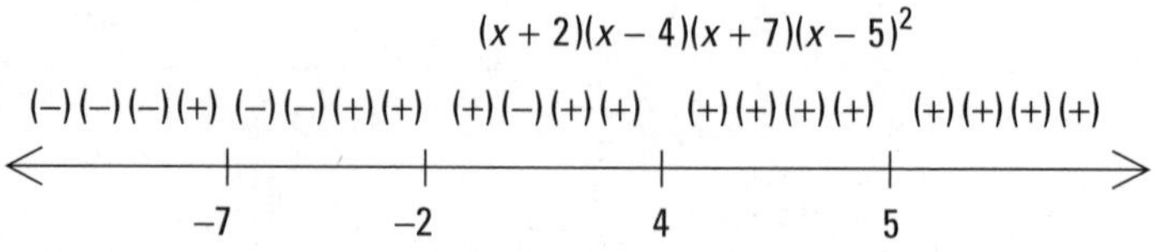

$(x+2)(x-4)(x+7)(x-5)^2$

(–)(–)(–)(+) (–)(–)(+)(+) (+)(–)(+)(+) (+)(+)(+)(+) (+)(+)(+)(+)

–7 –2 4 5

You want the expression on the left to be positive or 0, given the original language of the inequality. You find an even number of positive factors between –7 and –2 and for numbers greater than 4. You include the zeros, so the solution you find is $-7 \leq x \leq -2$ or $x \geq 4$. In interval notation, you write the solution as $[-7, -2]$ or $[4, \infty)$.

Chapter 4

Rolling Along with Rational and Radical Equations

In This Chapter

- Solving equations containing radicals and fractional exponents
- Working with negative exponents
- Recognizing quadratic-like equations and using unFOIL

Solving an algebraic equation requires some know-how. You need a game plan to solve equations with fractions, radicals, and negative or fractional exponents — one that involves careful planning and a final check of your answers. In this chapter, you find out how to tackle equations by changing them into new versions that are more familiar and easier to solve. You also see a recurring theme of *check your answers,* because changing equations into different forms can introduce mysterious strangers into the mix — in the form of false answers.

Rounding Up Rational Equations and Eliminating Fractions

An equation with one or more terms, at least one of which is rational, is called a *rational equation.* You probably hope that all your problems (and the people you associate with) are rational, but an equation that contains fractions isn't always easy to handle.

A general plan for solving a rational equation is to get rid of the fraction or fractions by changing the equation into an equivalent form with the same answer — a form that makes it easier to solve.

Two of the most common ways to get rid of the fractions are multiplying through by the least common denominator (LCD) and cross-multiplying proportions. I just happen to discuss both of these techniques in the sections that follow.

This mathematical sleight of hand — using alternate equations to solve more complicated problems — isn't without its potential problems. At times, the new equation produces an *extraneous solution* (also referred to as an *extraneous root*), a false solution that pops up because you messed around with the original format of the equation. To guard against including extraneous solutions in your answers, you need to check the solutions you come up with in the original equations.

Making your least common denominator work for you

You can solve many rational equations by simply getting rid of all the denominators (which gets rid of the fractions). To do so, you introduce the LCD into the problem. The LCD is the smallest number that all the denominators in the problem divide into evenly (such as 2, 3, and 4 all dividing the LCD 12 evenly).

To solve an equation using the LCD, you find the common denominator, write each fraction with that common denominator, and then multiply each side of the equation by that same denominator to get a nice fraction-less equation. The new equation is in an easier form to solve. I'll show you the step-by-step process with this example:

Solve for x in $\frac{3x+2}{2} - \frac{5}{2x-3} = \frac{x+3}{4}$.

1. **Find a common denominator.**

   The LCD is a multiple of each of the original denominators. To solve this equation, use $4(2x-3)$ as the LCD. All three denominators divide this product evenly.

2. **Write each fraction with the common denominator.**

   Multiply each fraction by the equivalent of 1. The numerator and denominator are the same, and the denominator is what is needed to change the original denominator into the LCD:

$$\frac{3x+2}{2}\cdot\frac{2(2x-3)}{2(2x-3)}-\frac{5}{2x-3}\cdot\frac{4}{4}=\frac{x+3}{4}\cdot\frac{2x-3}{2x-3}$$

   Completing the multiplication:

$$\frac{2(3x+2)(2x-3)}{4(2x-3)}-\frac{20}{4(2x-3)}=\frac{(x+3)(2x-3)}{4(2x-3)}$$

3. **Multiply each side of the equation by that same denominator.**

   Multiply each term in the equation by the LCD; then reduce each term and get rid of the denominators:

$$\cancel{4(2x-3)}\cdot\frac{2(3x+2)(2x-3)}{\cancel{4(2x-3)}}-\cancel{4(2x-3)}\cdot\frac{20}{\cancel{4(2x-3)}}=$$

$$\cancel{4(2x-3)}\cdot\frac{(x+3)(2x-3)}{\cancel{4(2x-3)}}$$

$$2(3x+2)(2x-3)-20=$$

$$(x+3)(2x-3)$$

4. **Solve the new equation.**

   To solve the new quadratic equation, you multiply out the terms, simplify, and set the equation equal to 0:

$$2(3x+2)(2x-3)-20=(x+3)(2x-3)$$

$$12x^2-10x-12-20=2x^2+3x-9$$

$$10x^2-13x-23=0$$

   Now you find out if the quadratic equation factors. If it doesn't factor, you can resort to the quadratic formula; fortunately, that isn't necessary here. After factoring, you set each factor equal to 0 and solve for x:

$$10x^2-13x-23=0$$

$$(10x-23)(x+1)=0$$

$$10x-23=0,\ x=\frac{23}{10}$$

$$x+1=0,\ x=-1$$

You find two solutions for the quadratic equation: $x = \frac{23}{10}$ and $x = -1$.

5. **Check your answers to avoid extraneous solutions.**

   You now have to check to be sure that both your solutions work in the *original* equation. ***Remember:*** One or both may be extraneous solutions.

   Checking the original equation to see if the two solutions work, you first look at $x = -1$. Replace each x with -1:

$$\frac{3(-1)+2}{2} - \frac{5}{2(-1)-3} = \frac{(-1)+3}{4}$$
$$\frac{-3+2}{2} - \frac{5}{-5} = \frac{2}{4}$$
$$-\frac{1}{2} + 1 = \frac{1}{2}$$
$$\frac{1}{2} = \frac{1}{2}$$

   Nice! The first solution works. The next check is to see if $x = \frac{23}{10}$ is a solution. And right now I'm going to take "author's privilege" and tell you that yes, the answer works. It takes more space than I have to show you all the steps, so I'm going to ask you to trust me and skip all the gory details. The two solutions of the rational equation are $x = -1$ and $x = \frac{23}{10}$.

Proposing proportions for solving rational equations

A *proportion* is an equation in which one fraction is set equal to another. Proportions have several very nice features that make them desirable to work with when you're solving rational equations because you can eliminate the fractions or change them so that they feature better denominators. Also, they factor in four different ways.

When you have the proportion $\frac{a}{b} = \frac{c}{d}$, the following are also true:

- ad and bc, the cross-products, are equal, giving you $ad = bc$.

- $\frac{b}{a}$ and $\frac{d}{c}$, the reciprocals, are equal, giving $\frac{b}{a} = \frac{d}{c}$.
- You can divide out common factors both horizontally and vertically.

Solve for x in the proportion: $\frac{80x}{16} = \frac{30}{x-5}$.

First reduce across the numerators, and then reduce the left fraction:

$$\frac{{}^{8}\cancel{80}x}{16} = \frac{\cancel{30}^{3}}{x-5} \text{ becomes } \frac{8x}{16} = \frac{3}{x-5}$$

$$\frac{{}^{1}\cancel{8}x}{{}_{2}\cancel{16}} = \frac{3}{x-5} \text{ becomes } \frac{x}{2} = \frac{3}{x-5}$$

Now cross-multiply and solve the resulting quadratic equation:

$$\begin{aligned} x(x-5) &= 6 \\ x^2 - 5x &= 6 \\ x^2 - 5x - 6 &= 0 \\ (x-6)(x+1) &= 0 \\ x = 6 &\text{ or } x = -1 \end{aligned}$$

As usual, you need to check to be sure that you haven't introduced any extraneous roots. Both solutions work!

Reasoning with Radicals

A radical in an equation often indicates that you want to find a root — the square root of a number, its cube root, and so on. A radical (root) adds a whole new dimension to what could've been a perfectly nice equation to solve. In general, you deal with radicals in equations the same way you deal with fractions in equations — you get rid of them. But watch out: extraneous answers often crop up in your work, so you have to check your answers.

Squaring both sides of the equation

If you have an equation with a square root term in it, you square both sides of the equation to get rid of the radical.

Solve for x in $\sqrt{4x+21}-6=x$.

First, add 6 to both sides of the equation to get the radical by itself on the left. Then square both sides of the equation.

$$\begin{aligned}\sqrt{4x+21}&=x+6\\ \left(\sqrt{4x+21}\right)^2&=(x+6)^2\\ 4x+21&=x^2+12x+36\end{aligned}$$

Now set the quadratic equation equal to 0 and solve it:

$$\begin{aligned}4x+21&=x^2+12x+36\\ 0&=x^2+8x+15\\ 0&=(x+3)(x+5)\\ x&=-3 \text{ or } x=-5\end{aligned}$$

The two solutions work for the quadratic equation that was created, but they don't necessarily work in the original equation. Check the work!

When $x=-3$, you get

$$\begin{aligned}\sqrt{4(-3)+21}-6&=\sqrt{-12+21}-6\\ &=\sqrt{9}-6\\ &=3-6\\ &=-3\end{aligned}$$

The solution $x=-3$ works. Checking $x=-5$, you get

$$\begin{aligned}\sqrt{4(-5)+21}-6&=\sqrt{-20+21}-6\\ &=\sqrt{1}-6\\ &=-5\end{aligned}$$

This solution works, too.

Both solutions working out is more the exception rather than the rule. Most of the time, one solution or the other works, but not both. And, unfortunately, sometimes you go through all the calculations and find that neither solution works in the original equation. You get an answer, of course (that there is no answer), but it isn't very fulfilling.

Taking on two radicals

Some equations that contain radicals call for more than one application of squaring both sides. For example, you have to square both sides more than once when you can't isolate a radical term by itself on one side of the equation. And you usually need to square both sides more than once when you have three terms in the equation — two of them with radicals.

Solve $\sqrt{3x+19}-\sqrt{5x-1}=2$.

1. **Move the radicals so that only one appears on each side.**
2. **Square both sides of the equation.**

   After the first two steps, you have the following:

   $$\left(\sqrt{3x+19}\right)^2=\left(2+\sqrt{5x-1}\right)^2$$
   $$3x+19=4+4\sqrt{5x-1}+5x-1$$

3. **Move all the nonradical terms to the left and simplify.**

   This gives you the following:

   $$3x+19-4-5x+1=4\sqrt{5x-1}$$
   $$-2x+16=4\sqrt{5x-1}$$

4. **Make the job of squaring the binomial on the left easier by dividing each term by 2 — the common factor of all the terms on both sides. Then square both sides, simplify, set the quadratic equal to 0, and solve for x.**

$$
\begin{aligned}
-x+8 &= 2\sqrt{5x-1} \\
(-x+8)^2 &= \left(2\sqrt{5x-1}\right)^2 \\
x^2-16x+64 &= 4(5x-1) \\
x^2-16x+64 &= 20x-4 \\
x^2-36x+68 &= 0 \\
(x-2)(x-34) &= 0 \\
x = 2 &\text{ or } x = 34
\end{aligned}
$$

The two solutions you come up with are $x = 2$ and $x = 34$. Both have to be checked in the original equation. When $x = 2$,

$$
\begin{aligned}
\sqrt{3(2)+19} - \sqrt{5(2)-1} &= \sqrt{25} - \sqrt{9} \\
&= 5-3 \\
&= 2
\end{aligned}
$$

When $x = 34$,

$$
\begin{aligned}
\sqrt{3(34)+19} - \sqrt{5(34)-1} &= \sqrt{121} - \sqrt{169} \\
&= 11-13 \\
&= -2
\end{aligned}
$$

The solution $x = 2$ works. The other solution, $x = 34$, doesn't work in the equation. The number 34 is an extraneous solution.

Dealing with Negative Exponents

Equations with negative exponents offer some unique challenges. In general, negative exponents are easier to work with if they disappear. Yes, as wonderful as negative exponents are in the world of mathematics, solving equations that contain them is often easier if you can change the format to positive exponents and fractions and then deal with solving the fractional equations (as shown in the previous section). What I do in this section, though, is show you how to handle negative exponents without resorting to the fractions.

A common type of equation with negative exponents is one with a mixture of powers. I show you how to deal with these particular equations by factoring out a greatest common factor (GCF). Another common negative-exponent problem is one that's quadratic-like.

Factoring out a negative exponent as a greatest common factor

The next example shows you, step-by-step, how to deal with an equation with negative exponents that can be solved by factoring.

Solve $3x^{-3} - 5x^{-2} = 0$ for x.

1. **Factor out the GCF.**

   In this case, the GCF is x^{-3}:

   $x^{-3}(3 - 5x) = 0$

   Did you think the exponent of the GCF was –2? ***Remember:*** –3 is smaller than –2. When you factor out a GCF, you choose the smallest exponent out of all the choices and then divide each term by that common factor.

2. **Set each term in the factored form equal to 0 to solve for *x*.**

   You end up with:

   $$x^{-3}(3 - 5x) = 0$$

   $$x^{-3} = 0, \ \frac{1}{x^3} = 0$$

   $$3 - 5x = 0, \ x = \frac{3}{5}$$

   The first equation has no solution. The fraction with 1 in the numerator and x^3 in the denominator is never equal to 0. The only way a fraction is equal to 0 is if the numerator is 0 (and the denominator is some other number).

3. **Check your answers.**

   The only solution for this equation is $\frac{3}{5}$ — a perfectly dandy answer.

$$3\left(\frac{3}{5}\right)^{-3}-5\left(\frac{3}{5}\right)^{-2}=3\left(\frac{5}{3}\right)^{3}-5\left(\frac{5}{3}\right)^{2}$$
$$=3\left(\frac{125}{27}\right)-5\left(\frac{25}{9}\right)$$
$$=\frac{125}{9}-\frac{125}{9}$$
$$=0$$

Solving quadratic-like trinomials

Trinomials are expressions with three terms, with the highest term raised to the second degree, the expression is quadratic. You can simplify quadratic trinomials by factoring them into two binomial factors. (See Chapter 3 for details on factoring quadratic-like trinomials.)

Solve the trinomial equation $3x^{-2}+5x^{-1}-2=0$.

You find the quadratic-like pattern: $ax^{-2n}+bx^{-n}+c$. Factoring and setting the two factors equal to 0:

$$\left(3x^{-1}-1\right)\left(x^{-1}+2\right)=0$$
$$3x^{-1}-1=0,\ \frac{3}{x}=1,\ x=3$$
$$x^{-1}+2=0,\ \frac{1}{x}=-2,\ x=-\frac{1}{2}$$

You produce two solutions, and both work when substituted into the original equation.

Be careful when solving an equation containing negative exponents — when the equation involves taking an even root (square root, fourth root, and so on). Watch out for zeros in the denominator, because those numbers don't exist, and be wary of imaginary numbers — they exist somewhere, in some mathematician's imagination. Factoring into binomials is a nifty way of solving equations with negative exponents — just be sure to proceed cautiously.

Fiddling with Fractional Exponents

You use fractional exponents ($x^{\frac{1}{2}}$, for example) to replace radicals and powers under radicals. Writing terms with fractional exponents allows you to perform operations on terms more easily when they have the same base or variable.

Solving equations by factoring fractional exponents

You can easily factor expressions that contain variables with fractional exponents if you know the rule for dividing numbers with the same base. To factor the expression $2x^{\frac{1}{2}} - 3x^{\frac{1}{3}}$, for example, you note that the smaller of the two exponents is the fraction $\frac{1}{3}$. Factor out x raised to that lower power, changing to a common denominator where necessary:

$$2x^{\frac{1}{2}} - 3x^{\frac{1}{3}} = x^{\frac{1}{3}}\left(2x^{\frac{1}{6}} - 3\right)$$

A good way to check your factoring work is to mentally distribute the first term through the terms in parentheses to be sure that the product is what you started with.

Promoting techniques for working with fractional exponents

Fractional exponents represent radicals and powers. Some equations with fractional exponents are solved by raising each side to an appropriate power to get rid of the fraction in the exponent. Other equations require various methods for solving equations, such as factoring.

Factoring out the greatest common factor

You don't always have the luxury of being able to raise each side of an equation to a power to get rid of the fractional exponents. Your next best plan of attack involves factoring out the variable with the smaller exponent and setting the two factors equal to 0.

Solve $x^{\frac{5}{6}} - 3x^{\frac{1}{2}} = 0$.

First factor out an x with the exponent of $\frac{1}{2}$. Then set the two factors equal to 0 to solve for x.

$$x^{\frac{5}{6}} - 3x^{\frac{1}{2}} = 0$$

$$x^{\frac{1}{2}}\left(x^{\frac{1}{3}} - 3\right) = 0$$

$$x^{\frac{1}{2}} = 0,\ x = 0$$

$$x^{\frac{1}{3}} - 3 = 0,\ x^{\frac{1}{3}} = 3,\ \left(x^{\frac{1}{3}}\right)^3 = (3)^3,\ x = 27$$

You come up with two perfectly civilized answers: $x = 0$ and $x = 27$.

Factoring quadratic-like fractional terms

Often, you can factor trinomials with fractional exponents into the product of two binomials. This is another version of the quadratic-like trinomials. After the factoring, you set the two binomials equal to 0 to determine if you can find any solutions.

Solve $x^{\frac{1}{2}} - 6x^{\frac{1}{4}} + 5 = 0$.

First, factor the left side into the product of two binomials. The exponent of the first term is twice that of the second, which should indicate to you that the trinomial has factoring potential. After you factor, you set the expression equal to 0 and solve for x:

$$\left(x^{\frac{1}{4}} - 1\right)\left(x^{\frac{1}{4}} - 5\right) = 0$$

$$x^{\frac{1}{4}} - 1 = 0,\ x^{\frac{1}{4}} = 1,\ \left(x^{\frac{1}{4}}\right)^4 = (1)^4,\ x = 1$$

$$x^{\frac{1}{4}} - 5 = 0,\ x^{\frac{1}{4}} = 5,\ \left(x^{\frac{1}{4}}\right)^4 = (5)^4,\ x = 625$$

Check your answers in the original equation; you find that both $x = 1$ and $x = 625$ work.

Chapter 5

Forging Function Facts

In This Chapter

- Defining functions, domain, and range
- Identifying one-to-one functions and even vs. odd functions
- Using function composition in the difference quotient

In algebra, the word *function* is very specific. You reserve it for certain math expressions that meet the tough standards of input and output values, as well as other mathematical rules of relationships. Therefore, when you hear that a certain relationship is a function, you know that the relationship meets some particular requirements. In this chapter, you find out more about these requirements. I also cover topics ranging from the domain and range of functions to the inverses of functions, and I show you how to perform the composition of functions. After acquainting yourself with these topics, you can confront a function equation with great confidence and a plan of attack.

Describing Function Characteristics

A *function* is a relationship between two variables that features exactly one output value for every input value — in other words, exactly one answer for every number inserted into the function rule.

For example, the equation $y = x^2 + 5x - 4$ is a function equation or function rule that uses the variables x and y. The x is the *input variable*, and the y is the *output variable*. If you input the number 3 for each of the x's, you get $y = 3^2 + 5(3) - 4 = 9 + 15 - 4 = 20$. The output is 20, the only possible answer. You won't get another number if you input the 3 a second time.

The single-output requirement for a function may seem like an easy requirement to meet, but you encounter plenty of strange math equations out there, so watch out.

Denoting function notation

Functions feature some special notation that makes working with them much easier. The notation doesn't change any of the properties — it just allows you to identify different functions quickly and indicate various operations and processes more efficiently.

The variables x and y are pretty standard in functions and come in handy when you're graphing functions. But mathematicians also use another format called *function notation.* For example, here are three particular functions named two different ways:

$$y = x^2 + 5x - 4 \qquad f(x) = x^2 + 5x - 4$$
$$y = \sqrt{3x - 8} \qquad g(x) = \sqrt{3x - 8}$$
$$y = 6xe^x - 2e^{2x} \qquad h(x) = 6xe^x - 2e^{2x}$$

On the left, you see the traditional x and y expression of the three functions. But when you see a bunch of functions written together, you can be efficient by referring to individual functions as f or g or h so listeners don't have to question what you're referring to. When I say, "Look at function g," your eyes go directly to the function I'm talking about.

Using function notation to evaluate functions

When you see a written function that uses function notation, you can easily identify the input variable, the output variable, and what you have to do to *evaluate* the function for some input (or replace the variables with numbers and simplify). You can do so because the input value is placed in the parentheses right after the function name or output value.

Evaluate $g(x) = \sqrt{3x - 8}$ when $x = 3$.

$g(3)$ is what you get when you substitute a 3 for every x in the function expression and perform the operations to get the output answer.

$g(3) = \sqrt{3(3) - 8} = \sqrt{9 - 8} = \sqrt{1} = 1$. Now you can say that $g(3) = 1$, or "g of 3 equals 1." The output of the function g is 1 if the input is 3.

Determining Domain and Range

The input and output values of a function (see the previous section) are of major interest to people working in algebra. The words *input* and *output* describe what's happening in the function (namely what number you put in and what result comes out), but the official designations for these sets of values are *domain* and *range.*

Delving into domain

The *domain* of a function consists of all the input values of the function. (Think of a king's domain of all his servants entering his kingdom.) In other words, the domain is the set of all numbers that you can input without creating an unwanted or impossible situation. Such situations can occur when operations appear in the definition of the function, such as fractions, radicals, logarithms, and so on.

Many functions have no exclusions of values, but fractions are notorious for causing trouble when zeros appear in the denominators. Radicals have restrictions as to what you can find roots of, and logarithms can only deal with positive numbers.

The way you express domain depends on what's required of the task you're working on — evaluating functions, graphing, determining a good fit as a model, to name a few. Here are some examples of functions and their respective domains:

- $\boldsymbol{f(x) = \sqrt{x - 11}}$**:** The domain consists of the number 11 and every greater number thereafter. You write this as $x \geq 11$ or, in interval notation, $[11, \infty)$. You can't use numbers smaller than 11 because you'd be taking the square root of a negative number, which isn't a real number.

- $g(x)=\frac{x}{x^2-4x-12}=\frac{x}{(x-6)(x+2)}$**:** The domain consists of all real numbers except 6 and –2. You write this domain as $x<-2$ or $-2<x<6$ or $x>6$, or, in interval notation, as $(-\infty, -2) \cup (-2, 6) \cup (6, \infty)$. It may be easier to simply write "All real numbers except $x=-2$ and $x=6$." The reason you can't use –2 or 6 is because these numbers result in a 0 in the denominator of the fraction, and a fraction with 0 in the denominator creates a number that doesn't exist.
- $h(x)=x^3-3x^2+2x-1$**:** The domain of this function is all real numbers. You don't have to eliminate anything, because you can't find a fraction with the potential of a zero in the denominator, and you have no radical to put a negative value into. You write this domain with a fancy R, $\Re$, or with interval notation as $(-\infty, \infty)$.

Wrangling with range

The *range* of a function is all its output values — all values you get by inputting the domain values into the *rule* (the function equation) for the function. You may be able to determine the range of a function from its equation, but sometimes you have to graph it to get a good idea of what's going on.

The following are some examples of functions and their ranges. Like domains (see the previous section), you can express ranges in words, inequalities, or interval notation:

- $k(x)=x^2+3$**:** The range of this function consists of the number 3 and any number greater than 3. You write the range as $k \geq 3$ or, in interval notation, $[3, \infty)$. The outputs can never be less than 3 because the numbers you input are squared. The result of squaring a real number is always positive (or if you input 0, you square 0). If you add a positive number or 0 to 3, you never get anything smaller than 3.
- $m(x)=\sqrt{x+7}$**:** The range of this function consists of all positive numbers and 0. You write the range as $m \geq 0$ or, in interval notation, $[0, \infty)$. The number under the radical can never be negative, and all the square roots come out positive or 0.

- $p(x) = \frac{2}{x-5}$: Some functions' equations, such as this one, don't give an immediate clue to the range values. It often helps to sketch the graphs of these functions. Figure 5-1 shows the graph of the function *p*. See if you can figure out the range values before peeking at the following explanation.

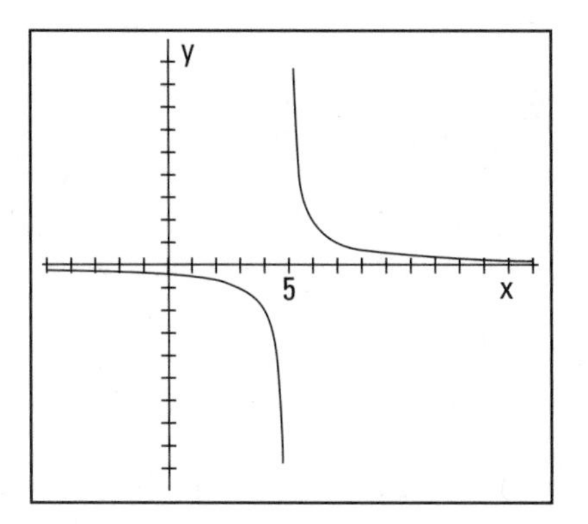

Figure 5-1: Try graphing equations that don't give an obvious range.

The graph of this function never touches the *x*-axis, but it gets very close. For the numbers in the domain bigger than 5, the graph has some really high *y* values and some *y* values that get really close to 0. But the graph never touches the *x*-axis, so the function value never really reaches 0. For numbers in the domain smaller than 5, the curve is below the *x*-axis. These function values are negative — some really small. But, again, the *y* values never reach 0. So, if you guessed that the range of the function is every real number except 0, you're right! You write the range as $p \neq 0$, or $(-\infty, 0) \cup (0, \infty)$. Did you also notice that the function doesn't have a value when $x = 5$? This happens because 5 isn't in the domain.

Counting on Even and Odd Functions

You can classify numbers as even or odd (and you can use this information to your advantage; for example, you know you can divide even numbers by 2 and come out with an integer). You can also classify some functions as even or odd.

Determining whether even or odd

An *even function* is one in which a domain value (an input) and its opposite always result in the same range value (output): $f(-x) = f(x)$ for every x in the domain. An *odd function* is one in which each domain value and its opposite produce opposite results in the range: $f(-x) = -f(x)$.

To determine if a function is even or odd (or neither), you replace every x in the function equation with $-x$ and simplify. If the function is even, the simplified version looks exactly like the original. If the function is odd, the simplified version looks like what you get after multiplying the original function equation by –1.

Show that $f(x) = x^4 - 3x^2 + 6$ is even.

Whether you input 2 or –2, you get the same output:

- $f(2) = (2)^4 - 3(2)^2 + 6 = 16 - 12 + 6 = 10$
- $f(-2) = (-2)^4 - 3(-2)^2 + 6 = 16 - 3(4) + 6 = 10$

So, you can say $f(2) = f(-2)$.

The example doesn't constitute a *proof* that the function is even; this is just a demonstration.

Show that $g(x) = x^3 - x$ is odd.

The inputs 2 and –2 give you opposite answers:

- $g(2) = (2)^3 - 2 = 8 - 2 = 6$
- $g(-2) = (-2)^3 - (-2) = -8 + 2 = -6$

So, you can say that $g(-2) = -g(2)$.

Again, I've *demonstrated,* not proved, that the function is odd.

You can't say that a function is even just because it has even exponents and coefficients, and you can't say that a function is odd just because the exponents and coefficients are odd numbers. If you do make these assumptions, you classify the functions incorrectly, which messes up your graphing. You have to apply the definitions to determine which label a function has.

Using even and odd functions in graphs

The biggest distinction of even and odd functions is how their graphs look:

- **Even functions:** The graphs of even functions are symmetric with respect to the *y*-axis (the vertical axis). You see what appears to be a mirror image to the left and right of the vertical axis. For an example of this type of symmetry, see Figure 5-2a, which is the graph of the even function $f(x) = \frac{5}{x^2+1}$.
- **Odd functions:** The graphs of odd functions are symmetric with respect to the origin. With this symmetry it looks the same if you rotate the graph by 180 degrees. The graph in Figure 5-2b, which is the odd function $g(x) = x^3 - 8x$, displays origin symmetry.

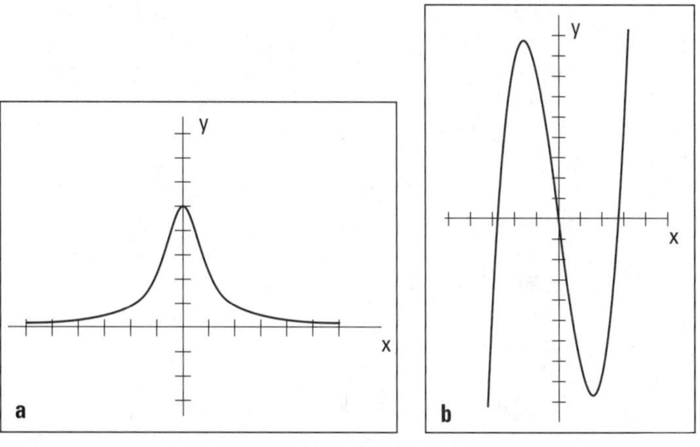

Figure 5-2: An even (a) and odd (b) function.

Taking on Functions One-to-One

Functions can have many classifications or names, depending on the situation and what you want to do with them. One very important classification is deciding whether a function is one-to-one.

Defining which functions are one-to-one

A function is *one-to-one* if you have exactly one output value for every input value *and* exactly one input value for every output value. Formally, you write this definition as follows:

> If f is a one-to-one function, then when $f(x_1) = f(x_2)$, it must be true that $x_1 = x_2$.
>
> In simple terms, if the two output values are the same, the two input values must also be the same.

One-to-one functions are important because they're the only functions that can have inverses, and functions with inverses aren't all that easy to come by. If a function has an inverse, you can work backward and forward — find an answer if you have a question and find the original question if you know the answer (sort of like *Jeopardy!*).

An example of a one-to-one function is $f(x) = x^3$. The rule for the function involves cubing the variable. The cube of a positive number is positive, and the cube of a negative number is negative. Therefore, every input has a unique output — no other input value gives you that output.

Some functions without the one-to-one designation may look like the previous example, which *is* one-to-one. Take $g(x) = x^3 - x$, for example. This counts as a function because only one output comes with every input. However, the function isn't one-to-one, because you can create the same output or function value from more than one input. For example, $g(1) = (1)^3 - (1) = 1 - 1 = 0$, and $g(-1) = (-1)^3 - (-1) = -1 + 1 = 0$. You have two inputs, 1 and –1, that result in the same output of 0.

Testing for one-to-one functions

You can determine which functions are one-to-one and which are violators by *sleuthing* (guessing and trying), using algebraic techniques and graphing. Most mathematicians prefer the graphing technique because it gives you a nice, visual answer. The basic graphing technique is the *horizontal line test.*

With the horizontal line test, you can see if any horizontal line drawn through the graph cuts through the function more than one time. If a line passes through the graph more than once, the function fails the test and, therefore, isn't a one-to-one function. Figure 5-3 shows a function that passes the horizontal line test and a function that flunks it.

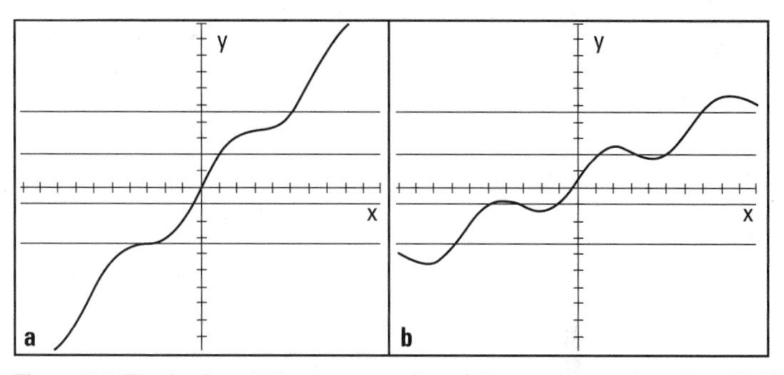

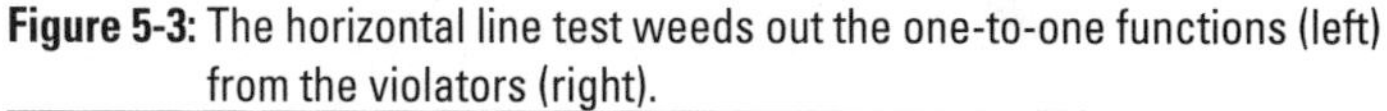

Figure 5-3: The horizontal line test weeds out the one-to-one functions (left) from the violators (right).

Composing Functions

You can perform the basic mathematical operations of addition, subtraction, multiplication, and division on the equations used to describe functions. For example, you can take the two functions $f(x) = x^2 - 3x - 4$ and $g(x) = x + 1$ and perform the four operations on them:

$$f + g = (x^2 - 3x - 4) + (x + 1) = x^2 - 2x - 3$$

$$f - g = (x^2 - 3x - 4) - (x + 1) = x^2 - 3x - 4 - x - 1 = x^2 - 4x - 5$$

$$f \cdot g = (x^2 - 3x - 4)(x + 1) = x^3 - 2x^2 - 7x - 4$$

$$\frac{f}{g} = \frac{x^2 - 3x - 4}{x + 1} = \frac{(x - 4)\cancel{(x + 1)}}{\cancel{x + 1}} = x - 4$$

Well done, but you have another operation at your disposal — an operation special to functions — called *composition*.

Composing yourself with functions

The *composition* of functions is an operation in which you use one function as the input into another and perform the operations on that input function.

You indicate the composition of functions f and g with a small circle between the function names, $(f \circ g)(x)$, and you define the composition as $(f \circ g)(x) = f(g(x))$.

Here's how you perform an example composition, using the functions $f(x) = x^2 - 3x - 4$ and $g(x) = x + 1$:

$$\begin{aligned}(f \circ g)(x) &= f(g(x)) \\ &= f(x+1) \\ &= (x+1)^2 - 3(x+1) - 4 \\ &= x^2 + 2x + 1 - 3x - 3 - 4 \\ &= x^2 - x - 6\end{aligned}$$

The composition of functions isn't commutative. (Addition and multiplication are *commutative,* because you can switch the order and not change the result.) The order in which you perform the composition — which function comes first — matters. The composition $(f \circ g)(x)$ isn't the same as $(g \circ f)(x)$.

Composing with the difference quotient

The *difference quotient* shows up in most high school algebra II classes as an exercise you do after your instructor shows you the composition of functions. You perform this exercise because the difference quotient is the basis of the definition of the derivative in calculus.

So, where does the composition of functions come in? With the difference quotient, you do the composition of some targeted function $f(x)$ and the function $g(x) = x + h$ or $g(x) = x + \Delta x$, depending on what calculus book you use.

The difference quotient for the function f is $\frac{f(x+h)-f(x)}{h}$.

Perform the difference quotient on $f(x) = x^2 - 3x - 4$.

$$\frac{f(x+h)-f(x)}{h} = \frac{\overbrace{(x+h)^2 - 3(x+h) - 4}^{f(x+h)} - \overbrace{\left(x^2 - 3x - 4\right)}^{f(x)}}{h}$$

Notice that you find the expression for $f(x + h)$ by putting $x + h$ in for every x in the function — $x + h$ is the input variable. Now, continuing on with the simplification:

$$= \frac{x^2 + 2xh + h^2 - 3x - 3h - 4 - x^2 + 3x + 4}{h}$$
$$= \frac{2xh + h^2 - 3h}{h}$$

Did you notice that x^2, $3x$, and 4 all appear in the numerator with their opposites? Now, to finish:

$$= \frac{h(2x + h - 3)}{h}$$
$$= 2x + h - 3$$

Now, this may not look like much to you, but you've created a wonderful result. You've just done some really decent algebra.

Getting Into Inverse Functions

Some functions are *inverses* of one another, but a function can have an inverse only if it's one-to-one. If two functions *are* inverses of one another, each function "undoes" what the other "did." In other words, you use them to get back where you started. The process is sort of like *Jeopardy!* — you have the answer and need to determine the question.

The notation for an inverse function is the exponent –1 written after the function name. The inverse of function $f(x)$, for example, is $f^{-1}(x)$.

Don't confuse the –1 exponent for taking the reciprocal of $f(x)$. The notation is what we're stuck with, so just pay heed.

Here are two inverse functions and how they can "undo" one another:

$$f(x) = \frac{x+3}{x-4} \text{ and } f^{-1}(x) = \frac{4x+3}{x-1}$$

$$f(5) = \frac{5+3}{5-4} = \frac{8}{1} = 8 \text{ and } f^{-1}(8) = \frac{4(8)+3}{8-1} = \frac{32+3}{7} = 5$$

If you put 5 into function f, you get 8 as a result. If you put 8 into f^{-1}, you get 5 as a result — you're back where you started.

Now, how can you tell when functions are inverses? Read on!

Finding which functions are inverses

In the example from the previous section, I tell you that two functions are inverses and then demonstrate how they work. You can't really *prove* that two functions are inverses by plugging in numbers, however. You may face a situation where a couple numbers work, but, in general, the two functions aren't really inverses.

The only way to be sure that two functions are inverses of one another is to use the following general definition:

> Functions f and f^{-1} are inverses of one another only if $f\left(f^{-1}(x)\right) = x$ *and* $f^{-1}\left(f(x)\right) = x$.

In other words, you have to do the composition in both directions and show that both result in the single value x.

Show that $f(x) = \sqrt[3]{2x-3} + 4$ and $g(x) = \frac{(x-4)^3 + 3}{2}$ are inverses of one another.

First, you perform the composition $f \circ g$:

$$
\begin{aligned}
f \circ g = f(g) &= \sqrt[3]{2(g)-3} + 4 \\
&= \sqrt[3]{\cancel{2}\left(\frac{(x-4)^3+3}{\cancel{2}}\right)-3} + 4 \\
&= \sqrt[3]{(x-4)^3+3-3} + 4 \\
&= \sqrt[3]{(x-4)^3} + 4 \\
&= (x-4)+4 = x
\end{aligned}
$$

Now you perform the composition in the opposite order:

$$
\begin{aligned}
g \circ f &= \frac{(f-4)^3+3}{2} \\
&= \frac{\left(\left(\sqrt[3]{2x-3}+4\right)-4\right)^3+3}{2} \\
&= \frac{\left(\sqrt[3]{2x-3}\right)^3+3}{2} \\
&= \frac{(2x-3)+3}{2} \\
&= \frac{2x}{2} \\
&= x
\end{aligned}
$$

Both come out with a result of x, so the functions are inverses of one another.

Finding an inverse of a function

Up until now in this section, I've given you two functions and told you that they're inverses of one another. I can show you how to create all sorts of inverses for all sorts of one-to-one functions.

Find the inverse of the one-to-one function $f(x) = \frac{x}{x-5}$.

1. **Rewrite the function, replacing $f(x)$ with y to simplify the notation.**

$$y = \frac{x}{x-5}$$

2. **Change each y to an x and each x to a y.**

$$x = \frac{y}{y-5}$$

3. **Solve for y.**

$$x = \frac{y}{y-5}$$
$$x(y-5) = y$$
$$xy - 5x = y$$
$$xy - y = 5x$$
$$y(x-1) = 5x,\ y = \frac{5x}{x-1}$$

4. **Rewrite the function, replacing the y with $f^{-1}(x)$.**

$$f^{-1}(x) = \frac{5x}{x-1}$$

Chapter 6

Graphing Linear and Quadratic Functions

In This Chapter

- Enlisting basic graphing techniques
- Identifying and graphing the equation of a line
- Graphing quadratic functions using vertices and intercepts

Graphing equations is an important part of understanding just what a function or other relationship represents. The modern, handheld graphing calculators take care of many of the details, but you still need to have a general idea of what the graph should look like so you know how to select a viewing window and so you know if you've entered something incorrectly.

Identifying Some Graphing Techniques

You do most graphing in algebra on the *Cartesian coordinate system* — a grid-like system where you plot points depending on the position and signs of numbers. Within the Cartesian coordinate system (which is named for the philosopher and mathematician Rene Descartes), you can plug-and-plot points to draw a curve, or you can take advantage of knowing a little something about what the graphs should look like. In either case, the coordinates and points fit together to give you a picture.

Graphing curves can take as long as you like or be as quick as you like. If you take advantage of the characteristics of the curves you're graphing, you can cut down on the time it takes to graph and improve your accuracy. Two features that you can quickly recognize and solve for are the intercepts and symmetry of the graphs.

Finding x- and y-intercepts

The *intercepts* of a graph appear at the points where the graph crosses the axes. The graph of a curve may never cross an axis, but when it does, knowing the points that represent the intercepts is very helpful.

The x-intercepts always have the format $(h, 0)$ — the y-coordinate is 0 because the point is on the x-axis. The y-intercepts have the form $(0, k)$ — the x-coordinate is 0 because the point is on the y-axis. You find the x- and y-intercepts by letting y and x, respectively, equal 0. To find the x-intercept(s) of a curve, you set y equal to 0 and solve a given equation for x. To find the y-intercept(s) of a curve, you set x equal to 0 and solve the equation for y.

Find the intercepts of the graph of $y = -x^2 + x + 6$.

To find the x-intercepts, let $y = 0$; you then have the quadratic equation $0 = -x^2 + x + 6 = -(x^2 - x - 6)$. Solve this equation by factoring it into $0 = -(x - 3)(x + 2)$. You find two solutions, $x = 3$ and -2, so the two x-intercepts are $(3, 0)$ and $(-2, 0)$. (For more on factoring, see Chapters 1 and 3.)

To find the y-intercept, let $x = 0$. This gives you the equation $y = -0 + 0 + 6 = 6$. The y-intercept, therefore, is $(0, 6)$.

Reflecting on a graph's symmetry

A graph that's *symmetric* with respect to one of the axes appears to be a mirror image of itself on either side of the axis. A graph symmetric about the origin appears to be the same after a 180-degree turn. Figure 6-1 shows three curves and three symmetries: symmetry with respect to the y-axis (a), symmetry with respect to the x-axis (b), and symmetry with respect to the origin (c).

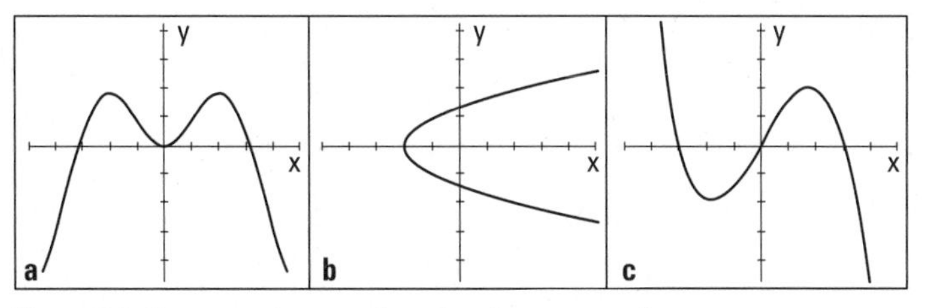

Figure 6-1: Symmetry in a graph makes for a pretty picture.

Recognizing that the graph of a curve has symmetry helps you sketch the graph and determine its characteristics. The following sections outline ways to determine, from a graph's equation, if symmetry exists.

With respect to the *y*-axis (even functions):

- If replacing every x with $-x$ doesn't change the value of y, the curve is the mirror image of itself over the *y*-axis. The graph contains the points (x, y) and $(-x, y)$.
- For example, the graph of the equation $y = x^4 - 3x^2 + 1$ is symmetric with respect to the *y*-axis. If you replace each x with $-x$, the equation remains unchanged. Replacing each x with $-x$, $y = (-x)^4 - 3(-x)^2 + 1 = x^4 - 3x^2 + 1$.

With respect to the *x*-axis:

- If replacing every y with $-y$ doesn't change the value of x, the curve is the mirror image of itself over the *x*-axis. The graph contains the points (x, y) and $(x, -y)$.
- For example, the graph of $x = \frac{10}{y^2+1}$ is symmetric with respect to the *x*-axis. When you replace each y with $-y$, the *x*-value remains unchanged.

With respect to the origin (odd functions):

- If replacing every variable with its opposite is the same as multiplying the entire equation by –1, the curve can rotate by 180 degrees about the origin and be its own image. The graph contains the points (x, y) and $(-x, -y)$.
- For example, the graph of $y = x^5 - 10x^3 + 9x$ is symmetric with respect to the origin. When you replace every x and y with $-x$ and $-y$, you get $-y = -x^5 + 10x^3 - 9x$, which is the same as multiplying everything through by –1.

Mastering the Graphs of Lines

Lines are some of the simplest graphs to sketch. It takes only two points to determine the one and only line that passes through them and goes on forever and ever in a space, so one simple method for graphing lines is to find two points — any two points — on the line. Another useful method is to use a point and the slope of the line. The method you choose is often just a matter of personal preference.

The slope of a line also plays a big role in comparing it with other lines that run parallel or perpendicular to it. The slopes are closely related to one another.

Determining the slope of a line

The *slope* of a line, designated by the letter m, has a complicated math definition, but it's basically a number — positive, negative, or zero; large or small — that tells you something about the steepness and direction of the line. The numerical value of the slope tells you if the line slowly rises or drops from left to right or dramatically soars or falls from left to right.

Characterizing a line's slope

A line can have a positive slope, a negative slope, a zero slope, or no slope at all. The greater the *absolute value* (the value of the number without regard to the sign; in other words, the distance of the number from 0) of a line's slope, the steeper the line is. For example, if the slope is a number between –1 and 1, the line is rather flat. A slope of 0 means that the line is absolutely horizontal.

A vertical line doesn't have a slope. This is tied to the fact that numbers go infinitely high, and math doesn't have a highest number — you just say *infinity.* Only an infinitely high number can represent a vertical line's slope, but usually, if you're talking about a vertical line, you just say that the slope doesn't exist.

Computing a line's slope

You can determine the slope of a line, m, if you know two points on the line.

You find the slope of the line that goes through the points (x_1, y_1) and (x_2, y_2) with the formula $m = \frac{y_2 - y_1}{x_2 - x_1}$.

Find the slope of the line through (–3, 2) and (4, –12).

Use the formula to get $m = \frac{y_2 - y_1}{x_2 - x_1} = \frac{-12 - 2}{4 - (-3)} = \frac{-14}{7} = -2$. This line is fairly steep — the absolute value of –2 is 2 — and it falls as it moves from left to right, which makes its slope negative.

When you use the slope formula, it doesn't matter which point you choose to be (x_1, y_1) — the order of the points doesn't matter — but you can't mix up the order of the two different coordinates. You can run a quick check by seeing if the coordinates of each point are above and below one another. Also, be sure that the *y*-coordinates are in the numerator; a common error is to have the difference of the y-coordinates in the denominator.

Describing two line equations

I offer two different forms for the equation of a line. The first is the *standard form,* written $Ax + By = C$, with the two variable terms on one side and the constant on the other side. The other form is the *slope-intercept form,* written $y = mx + b$; the *y*-value is set equal to the product of the slope, *m*, and *x* added to the *y*-intercept, *b*.

Standing up with the standard form

The standard form has more information about the line than may be immediately apparent. You can determine, just by looking at the numbers in the equation, the intercepts and slope of the line.

The line $Ax + By = C$ has

- An *x*-intercept of $\left(\frac{C}{A}, 0\right)$
- A *y*-intercept of $\left(0, \frac{C}{B}\right)$
- A slope of $m = -\frac{A}{B}$

Graph the line $4x + 3y = 12$ using the intercepts.

Plot the intercepts, $\left(\frac{C}{A},0\right)=\left(\frac{12}{4},0\right)=(3,0)$ and $\left(0,\frac{C}{B}\right)=\left(0,\frac{12}{3}\right)=(0,4)$.

Then draw the line through them. Figure 6-2 shows the two intercepts and the graph of the line. Note that the line falls as it moves from left to right, confirming the negative value of the slope from the formula $m=-\frac{A}{B}=-\frac{4}{3}$.

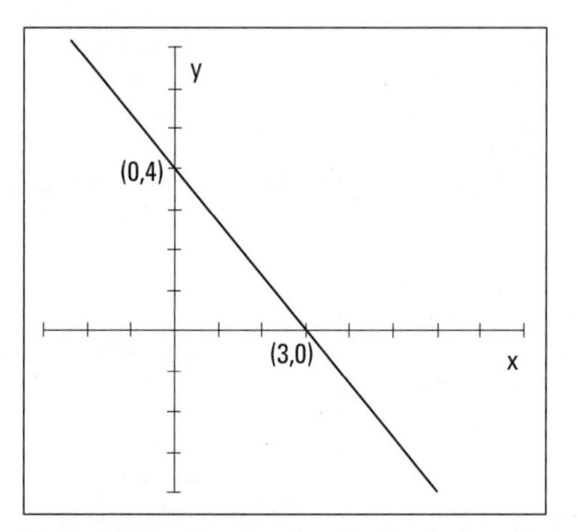

Figure 6-2: Graphing $4x+3y=12$, a line written in standard form, using its intercepts.

Sliding down the slope-intercept form

When the equation of a line is written in the slope-intercept form, $y = mx + b$, you have good information right at your fingertips. The coefficient of the x term, m, is the slope of the line. And the constant, b, is the y-value of the y-intercept. With these two bits of information, you can quickly sketch the line.

If you want to graph the line $y = 2x + 5$, for example, you first plot the y-intercept, (0, 5), and then "count off" the slope from that point moving to the right and then up or down. The slope of the line $y = 2x + 5$ is 2; think of the 2 as the slope fraction, with the y-coordinates on top and the x-coordinates on bottom. The slope then becomes $\frac{2}{1}$. So you move one unit to the right and then two units up, because the slope is positive.

Changing from one form to the other

You can graph lines by using the standard form or the slope-intercept form of the equations. If you prefer one form to the other — or if you need a particular form for an application you're working on — you can change the equations to your preferred form by performing simple algebra:

- To change the standard form to the slope-intercept form, you just solve for y.
- To change the slope-intercept form to the standard form, you rewrite the equation with the x and y terms on one side and then multiply through by a constant to create integer coefficients and a constant on the other side.

Identifying parallel and perpendicular lines

Lines are *parallel* when they never touch — no matter how far out you draw them. Lines are *perpendicular* when they intersect at a 90-degree angle. Both of these instances are fairly easy to spot when you see the lines graphed, but how can you be sure that the lines are truly parallel or that the angle is really 90 degrees and not 89.9 degrees? The answer lies in the slopes.

Consider two lines, $y = m_1x + b_1$ and $y = m_2x + b_2$.

Two lines are *parallel* when their slopes are equal ($m_1 = m_2$). Two lines are *perpendicular* when their slopes are negative reciprocals of one another: $\left(m_2 = -\frac{1}{m_1}\right)$.

For example, the lines $y = 3x + 7$ and $y = 3x - 2$ are parallel. Both lines have a slope of 3, but their y-intercepts are different — one crosses the y-axis at 7 and the other at –2. The lines $y = -\frac{3}{8}x + 4$ and $y = \frac{8}{3}x - 2$ are perpendicular. The slopes are negative reciprocals of one another.

Coming to Terms with the Standard Form of a Quadratic

A *parabola* is the graph of a quadratic function. The graph is a nice, gentle, U-shaped curve that has points located an equal distance on either side of a line running up through its middle — called its *axis of symmetry.* Parabolas can be turned upward, downward, left, or right, but parabolas that represent functions only turn up or down. Here's the standard form for the quadratic function:

$$f(x) = ax^2 + bx + c$$

The coefficients (multipliers of the variables) a, b, and c are real numbers; a can't be equal to 0 because you'd no longer have a quadratic function. There's meaning in everything — or nothing!

Starting with "a" in the standard form

As the lead coefficient of the standard form of the quadratic function $f(x) = ax^2 + bx + c$, a provides important information:

- If a is positive, the graph of the parabola opens upward.
- If a is negative, the graph of the parabola opens downward.
- If a has an absolute value greater than 1, the graph of the parabola is "steep."
- If a has an absolute value less than 1, the graph of the parabola flattens.

Figure 6-3 shows some representatives of the different directions and forms that parabolas can take.

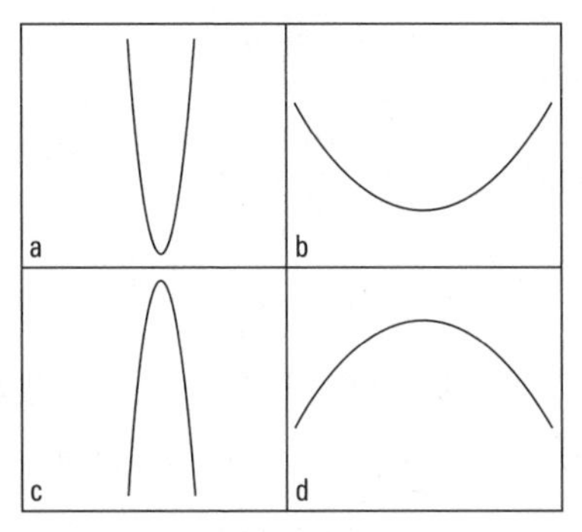

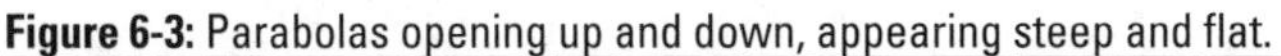
Figure 6-3: Parabolas opening up and down, appearing steep and flat.

The following equations of parabolas demonstrate for you the effect of the coefficient on the squared term:

- $y = 4x^2 - 3x + 2$**:** You say that this parabola is steep and opens upward because the lead coefficient is positive and greater than 1.
- $y = -\frac{1}{3}x^2 + x - 11$**:** You say that this parabola is flattened out and opens downward because the lead coefficient is negative, and the absolute value of the fraction is less than 1.
- $y = 0.002x^2 + 3$**:** You say that this parabola is flattened out and opens upward because the lead coefficient is positive, and the decimal value is less than 1. In fact, the coefficient is so small that the flattened parabola almost looks like a horizontal line.

Following "a" with "b" and "c"

Much like the lead coefficient in the quadratic function (see the previous section), the terms *b* and *c* give you plenty of information. Mainly, the terms tell you a lot if they're *not* there.

- If the second coefficient, b, is 0, the parabola straddles the y-axis. The parabola's *vertex* — the highest or lowest point on the curve, depending on which way it faces — is on that axis, and the parabola is symmetric about the axis. The equation then takes the form $y = ax^2 + c$.
- If the last coefficient, c, is 0, the graph of the parabola goes through the origin — in other words, one of its intercepts is the origin. The equation then becomes $y = ax^2 + bx$, which you can easily factor into $y = x(ax + b)$.

Eyeing a Quadratic's Intercepts

The *intercepts* of a quadratic function (or any function) are the points where the graph of the function crosses the x- or y-axis.

Intercepts are very helpful when you're graphing a parabola. The points are easy to find because one of the coordinates is always 0. If you have the intercepts and the vertex, and you use the symmetry of the parabola, you have a good idea of what the graph looks like.

Finding the one and only y-intercept

The y-intercept of a quadratic function is $(0, c)$. A parabola with the standard equation $y = ax^2 + bx + c$ is a function, so by definition, only one y-value can exist for every x-value. When $x = 0$, then $y = c$ and the y-intercept is $(0, c)$.

To find the y-intercepts of the following functions, you let $x = 0$:

- $\mathbf{y = 4x^2 - 3x + 2}$**:** When $x = 0$, $y = 2$ (or $c = 2$). The y-intercept is $(0, 2)$.
- $\mathbf{y = -x^2 - 5}$**:** When $x = 0$, $y = -5$ (or $c = -5$). Don't let the missing x term throw you. The y-intercept is $(0, -5)$.
- $\mathbf{y = x^2 + 9x}$**:** When $x = 0$, $y = 0$. The equation provides no constant term; you could also say the missing constant term is 0. The y-intercept is $(0, 0)$.

Getting at the x-intercepts

You find the x-intercepts of quadratics when you solve for the *zeros*, or solutions, of a quadratic equation and find real number answers. Parabolas with an equation of the standard form $y = ax^2 + bx + c$ open upward or downward and may or may not have x-intercepts; when the equation $0 = ax^2 + bx + c$ has no real solutions, then the graph has no x-intercepts.

The coordinates of all x-intercepts have zeros in them. An x-intercept's y-value is 0, and you write it in the form $(h, 0)$. How do you find the value of h? You let $y = 0$ in the general equation and then solve for x. You have two options to solve the equation $0 = ax^2 + bx + c$:

- Use the quadratic formula (see Chapter 3).
- Factor the expression and use the multiplication property of zero (MPZ; see Chapter 1).

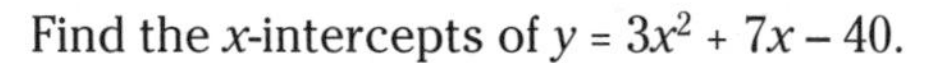

Find the x-intercepts of $y = 3x^2 + 7x - 40$.

Set y equal to 0 and solve the quadratic equation by factoring:

$$0 = 3x^2 + 7x - 40 = (3x - 8)(x + 5)$$

So $x = \frac{8}{3}$ or $x = -5$.

The two x-intercepts are $\left(\frac{8}{3}, 0\right)$ and $(-5, 0)$.

This next example shows how you determine that an equation has no x-intercept.

Find the x-intercepts of $y = -2x^2 + 4x - 7$.

Set y equal to 0 and you find that the quadratic doesn't factor. Then you apply the quadratic formula.

$$\begin{aligned} x &= \frac{-4 \pm \sqrt{4^2 - 4(-2)(-7)}}{2(-2)} \\ &= \frac{-4 \pm \sqrt{16 - (56)}}{-4} \\ &= \frac{-4 \pm \sqrt{-40}}{-4} \end{aligned}$$

You see that the value under the radical is negative; there are no real solutions. Alas, you find no x-intercept for this parabola.

Finding the Vertex of a Parabola

Quadratic functions, or parabolas, that have the standard form $y = ax^2 + bx + c$ are gentle, U-shaped curves that open either upward or downward. When the lead coefficient, a, is a positive number, the parabola opens upward, creating a *minimum value* for the function — the function values never go lower than that minimum. When a is negative, the parabola opens downward, creating a *maximum value* for the function — the function values never go higher than that maximum. The two extreme values, the minimum and maximum, occur at the parabola's *vertex*. The y-coordinate of the vertex gives you the numerical value of the extreme — its highest or lowest point. And the x-coordinate is part of the equation of the axis of symmetry.

Computing vertex coordinates

Finding the vertex of the parabola representing a quadratic function is as easy as a, b, c — without the c. Just insert the coefficients a and b into a formula.

The parabola $y = ax^2 + bx + c$ has its vertex when the x-value is equal to $\frac{-b}{2a}$. You plug in the a and b values from the equation to come up with the x-coordinate, and then you find the y-coordinate of the vertex by plugging this x-value into the equation and solving for y.

Find the coordinates of the vertex of $y = -3x^2 + 12x - 7$.

Solving for x, use the coefficients a and b:

$$x = \frac{-12}{2(-3)} = \frac{-12}{-6} = 2$$

You solve for y by putting the x-value back into the equation:

$$y = -3(2)^2 + 12(2) - 7 = -12 + 24 - 7 = 5$$

The coordinates of the vertex are (2, 5). This is the highest point for the parabola, because a is a negative number, which means the parabola opens downward from this point.

Linking up with the axis of symmetry

The *axis of symmetry* of a quadratic function is a vertical line that runs through the vertex of the parabola and acts as a mirror — half the parabola rests on one side of the axis, and half rests on the other. The x-value in the coordinates of the vertex appears in the equation for the axis of symmetry. For example, if a vertex has the coordinates (2, 3), the axis of symmetry is $x = 2$. All vertical lines have an equation of the form $x = h$. In the case of the axis of symmetry, the h is always the x-coordinate of the vertex.

Sketching a Graph from the Available Information

You have all sorts of information available when it comes to a quadratic function and its graph. You can use the intercepts, the opening, the steepness, the vertex, the axis of symmetry, or just some random points to plot the parabola. You don't really need all the pieces for each graph; as you practice sketching these curves, it becomes easier to figure out which pieces you need for different situations. The example I give, though, will use all the different possibilities — and each will just verify all the others.

Sketch the graph of $y = x^2 - 4x - 5$.

First, notice that the equation represents a parabola that opens upward, because the lead coefficient, a, is positive (+1). The y-intercept is (0, –5), which you get by plugging in 0 for x. If you set y equal to 0 to solve for the x-intercepts, you get $0 = x^2 - 4x - 5$, which factors into $0 = (x + 1)(x - 5)$. The x-intercepts are (–1, 0) and (5, 0).

The vertex is found using the formula for the x-coordinate of the vertex to get $x = \frac{-(-4)}{2(1)} = 2$. Plug the 2 into the formula for the parabola, and you find that the vertex is at (2, –9).

Use the axis of symmetry, which is $x = 2$, to find some points on either side — to help you with the shape. If you let $x = 6$, for example, you find that $y = 7$. This point is four units to the right of $x = 2$; four units to the left of $x = 2$ is $x = -2$. The corresponding point is found by putting –2 into the equation for the parabola; you get (–2, 7).

Use all that information in a graph to produce a sketch of the parabola (see Figure 6-4).

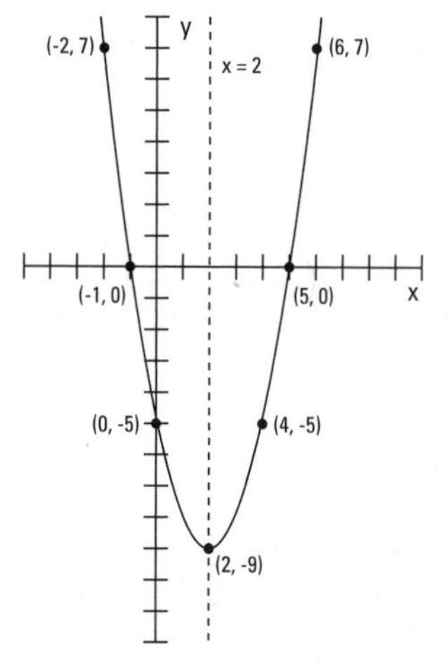

Figure 6-4: Using the various pieces of a quadratic as steps for sketching a graph ($y = x^2 - 4x - 5$).

Chapter 7

Pondering Polynomials

In This Chapter

- Providing techniques for making graphing polynomials easier
- Segueing from intercepts to roots of polynomials
- Solving polynomial equations using everything but the kitchen sink

The word *polynomial* comes from *poly-*, meaning "many," and *-nomial,* meaning "name" or "designation." The exponents used in polynomials are all whole numbers — no fractions or negatives. Polynomials get progressively more interesting as the exponents get larger — they can have more intercepts and turning points. This chapter outlines how to deal with polynomials: factoring them, graphing them, analyzing them. The graph of a polynomial looks like a Wisconsin landscape — smooth, rolling curves. Are you ready for this ride?

Sizing Up a Polynomial Equation

A *polynomial function* is a specific type of function that can be easily spotted in a crowd of other types of functions and equations. By convention, you write the terms from the largest exponent to the smallest.

The general form for a polynomial function is

$$f(x) = a_n x^n + a_{n-1}x^{n-1} + a_{n-2}x^{n-2} + \ldots + a_1 x^1 + a_0$$

Here, the a's are real numbers and the n's are whole numbers. The last term is technically $a_0 x^0$, if you want to show the variable in every term.

Identifying Intercepts and Turning Points

The *intercepts* of a polynomial are the points where the graph of the curve of the polynomial crosses the x-axis and y-axis. A polynomial function has *exactly* one y-intercept, but it can have many x-intercepts, depending on the degree of the polynomial (the highest power of the variable). The higher the degree, the more x-intercepts are possible.

The *x-intercepts* of a polynomial are also called the *roots, zeros,* or *solutions.* The x-intercepts are often where the graph of the polynomial goes from positive values (above the x-axis) to negative values (below the x-axis) or from negative values to positive values. Sometimes, though, the values on the graph don't change sign at an x-intercept: These graphs look sort of like a *touch and go.* The curves approach the x-axis, seem to change their minds about crossing the axis, touch down at the intercepts, and then go back to the same side of the axis.

A *turning point* of a polynomial is where the graph of the curve changes direction. It can change from going upward to going downward, or vice versa. A turning point is where you find a maximum value of the polynomial or a minimum value.

Interpreting relative value and absolute value

A parabola opening downward has an absolute maximum — you see no point on the curve that's higher than the maximum. In other words, no value of the function is greater than the function value at that point. Some functions, however, also have *relative* maximum or minimum values:

- **Relative maximum:** A function value that is bigger than all function values around it — it's *relatively* large. The function value is bigger than anything around it, but you may be able to find a bigger function value somewhere else.
- **Relative minimum:** A function value that is smaller than all function values around it. The function value is smaller than anything close to it, but there may be a function value that's smaller somewhere else.

In Figure 7-1, you can see five turning points. Two correspond to relative maximum values, which means they're higher than any points close to them. Three correspond to minimum values, which means they're lower than any points around them. Two of the minimums correspond to relative minimum values, and one has absolutely the lowest function value on the curve. This function has no absolute maximum value because it keeps going up and up without end.

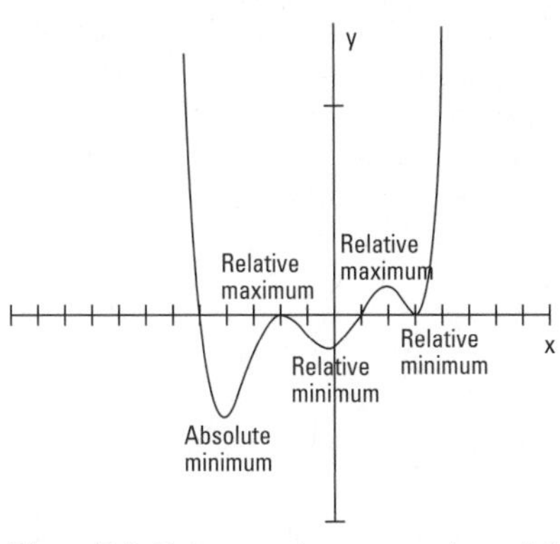

Figure 7-1: Extreme points on a polynomial.

Dealing with intercepts and turning points

The number of potential turning points and *x*-intercepts of a polynomial function is good to know when you're sketching the graph of the function. You can often count the number of *x*-intercepts and turning points of a polynomial if you have the graph of it in front of you, but you can also make an estimate of the number if you have the equation of the polynomial. Your estimate is actually a number that represents the most points that can occur.

Given the polynomial $f(x) = a_n x^n + a_{n-1}x^{n-1} + a_{n-2}x^{n-2} + \ldots + a_1 x^1 + a_0$, the maximum number of *x*-intercepts is *n*, the degree or highest power of the polynomial. The maximum number of turning points is $n - 1$, or one less than the number of possible intercepts. You may find fewer *x*-intercepts than *n*, or you may find exactly that many.

Examine the function equations for intercepts and turning points:

$$f(x) = 2x^7 + 9x^6 - 75x^5 - 317x^4 + 705x^3 + 2{,}700x^2$$

This graph has at most seven x-intercepts (7 is the highest power in the function) and six turning points (7 – 1).

You can see the graph of the function in Figure 7-2. According to its equation, the graph of the polynomial could have as many as seven x-intercepts, but it has only five; it does have all six turning points, though. You can also see that two of the intercepts are touch-and-go types, meaning that they approach the x-axis before heading away again.

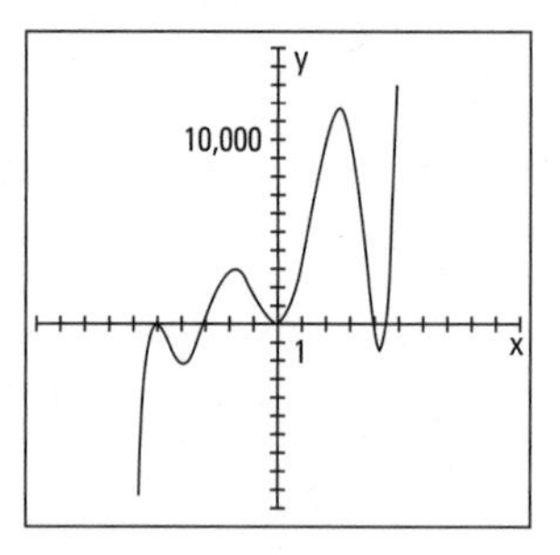

Figure 7-2: The intercept and turning-point behavior of a polynomial function.

Solving for y-intercepts and x-intercepts

You can easily solve for the y-intercept of a polynomial function; the y-intercept is where the curve of the graph crosses the y-axis, and that's when $x = 0$. So, to determine the y-intercept for any polynomial, simply replace all the x's with zeros and solve for y (that's the y part of the coordinates of that intercept). For example, in $y = 3x^4 - 2x^2 + 5x - 3$, you get $y = 3(0)^4 - 2(0)^2 + 5(0) - 3 = -3$, so the y-intercept is $(0, -3)$.

After you complete the easy task of solving for the y-intercept, you find out that the x-intercepts are another matter altogether. The value of y is 0 for all x-intercepts, so you let $y = 0$ and solve.

When the polynomial is factorable, you use the multiplication property of zero (MPZ; see Chapter 1), setting the factored form equal to 0 to find the x-intercepts.

Determine the x-intercepts of the polynomial $y = x^3 - 16x$.

Replace the y with zeros and solve for x:

$$0 = x^3 - 16x = x(x^2 - 16) = x(x - 4)(x + 4)$$

Using the MPZ, you get that $x = 0$, $x = 4$, or $x = -4$. The x-intercepts are (0, 0), (4, 0), and (–4, 0).

Determining When a Polynomial Is Positive or Negative

When a polynomial has positive y-values for some interval — between two x-values — its graph lies above the x-axis in that interval. When a polynomial has negative values, its graph lies below the x-axis in that interval. The only way for a polynomial to change from positive to negative values or vice versa is to go through 0 — at an x-intercept.

Incorporating a sign line

If you're a visual person like me, you'll appreciate the interval method I present in this section. Using a *sign line* and marking the intervals between x-values allows you to determine where a polynomial is positive or negative, and it appeals to your artistic bent!

Determine when the function $f(x) = x(x - 2)(x - 7)(x + 3)$ is positive and when it's negative.

Setting $f(x) = 0$ and solving, you find that the x-intercepts are at $x = 0, 2, 7$, and -3. To determine the positive and negative intervals for a polynomial function, follow this method:

1. **Draw a number line, and place the values of the x-intercepts in their correct positions on the line.**

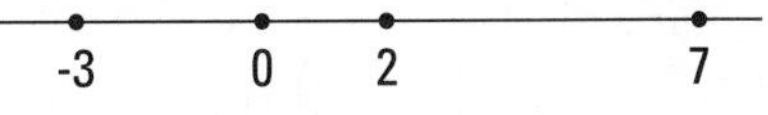

2. **Choose random values to the right of and left of and in between the intercepts to test whether the function is positive or negative in those intervals.**

   One efficient method is to insert the "test values" into the factored form of the polynomial and just record the signs — which then give you the positive or negative result for the entire interval.

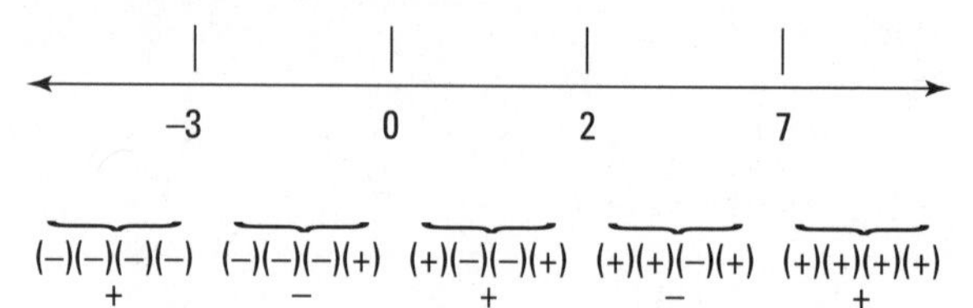

   You need to check only one point in each interval; the function values all have the same sign within that interval.

   The graph of this function is positive, or above the x-axis, whenever x is smaller than –3, between 0 and 2, or bigger than 7. You write this part of the answer as: $x < -3$ or $0 < x < 2$ or $x > 7$. The graph of the function is negative when $-3 < x < 0$ or $2 < x < 7$.

Recognizing a sign change rule

In the previous example, you see the signs changing at each intercept. If the signs of functions don't change at an intercept, then the graph of the polynomial doesn't cross the x-axis at that intercept, and you see a touch-and-go. It's nice to be able to predict such behavior.

The rule for whether a function displays sign changes or not at the intercepts is based on the exponent on the factor that provides you with a particular intercept.

If a polynomial function is factored in the form $y = (x - a_1)^{n_1}(x - a_2)^{n_2}\ldots$, you see a sign change at a_1 whenever n_1 is an odd number (meaning it crosses the x-axis), and you see no sign change whenever n_1 is even (meaning the graph of the function is touch-and-go; see the "Dealing with intercepts and turning points" section, earlier in this chapter).

So, for example, with the function $y = x^4(x - 3)^3(x + 2)^8(x + 5)^2$, you'll find a sign change at $x = 3$ and no sign change at $x = 0$, -2, or -5. And with the function $y = (2 - x)^2(4 - x)^2(6 - x)^2(2 + x)^2$, you never see a sign change — the function is always either positive or just touching the x-axis.

Solving Polynomial Equations

Finding intercepts (or roots or zeros) of polynomials can be relatively easy or a little challenging, depending on the complexity of the function. Polynomials that factor easily are very desirable. Polynomials that don't factor at all, however, are relegated to computers or graphing calculators.

The polynomials that remain are those that factor — but take a little planning and work. The planning process involves counting the number of possible positive and negative real roots and making a list of potential rational roots. The work is done using synthetic division to test the list of choices to find the roots.

Factoring for roots

Finding x-intercepts of polynomials isn't difficult — as long as you have the polynomial in nicely factored form. You just set the y equal to 0 and use the MPZ. This section deals with easily recognizable factors of polynomials; I cover other, more challenging types in the following sections.

Half the battle when factoring is recognizing the patterns in factorable polynomial functions. Here are the most easily recognizable factoring patterns used on polynomials:

- **Difference of squares:** $a^2 - b^2 = (a + b)(a - b)$.
- **Greatest common factor (GCF):** $ab \pm ac = a(b \pm c)$.
- **Difference of cubes:** $a^3 - b^3 = (a - b)(a^2 + ab + b^2)$.
- **Sum of cubes:** $a^3 + b^3 = (a + b)(a^2 - ab + b^2)$.
- **Perfect square trinomial:** $a^2 \pm 2ab + b^2 = (a \pm b)^2$.
- **Trinomial factorization:** UnFOIL (see Chapter 1).
- **Common factors in groups:** Grouping (see Chapter 1).

The following examples incorporate the different methods of factoring. They contain perfect cubes and squares and all sorts of good combinations of factorization patterns.

Factor the polynomial: $y = 4x^5 - 25x^3$.

First use the GCF and then the difference of squares:

$$y = 4x^5 - 25x^3 = x^3(4x^2 - 25) = x^3(2x - 5)(2x + 5)$$

Factor the polynomial: $y = 64x^8 - 64x^6 - x^2 + 1$.

You initially factor the polynomial by grouping. The first two terms have a common factor of $64x^6$, and the second two terms have a common factor of –1. The new equation has a common factor of $x^2 - 1$. After performing the factorization, you see that both factors are the difference of squares:

$$\begin{aligned} y &= 64x^8 - 64x^6 - x^2 + 1 \\ &= 64x^6\left(x^2 - 1\right) - 1\left(x^2 - 1\right) \\ &= \left(x^2 - 1\right)\left(64x^6 - 1\right) \end{aligned}$$

Now you factor the binomials as the difference of perfect squares. Then you can factor the last two new binomials using the difference and sum of two perfect cubes:

$$\begin{aligned} &= (x - 1)(x + 1)\left(8x^3 - 1\right)\left(8x^3 + 1\right) \\ &= (x - 1)(x + 1)(2x - 1)\left(4x^2 + 2x + 1\right)(2x + 1)\left(4x^2 - 2x + 1\right) \end{aligned}$$

The two trinomials resulting from factoring the difference and sum of cubes don't factor, so you're done. Whew!

Taking sane steps with the rational root theorem

What do you do if the factorization of a polynomial doesn't leap out at you? You have a feeling that the polynomial factors, but the necessary numbers escape you. Never fear! The rational root theorem is here.

The *rational root theorem* states that if the polynomial $f(x) = a_n x^n + a_{n-1}x^{n-1} + a_{n-2}x^{n-2} + \ldots + a_1 x^1 + a_0$ has any rational roots, they all meet the requirement that you can write them as a fraction equal to $\frac{\text{factor of } a_0}{\text{factor of } a_n}$.

In other words, according to the theorem, any rational root of a polynomial with integer coefficients is formed by dividing a factor of the constant term by a factor of the lead coefficient. Of course, this means that the a_0 term, the constant, cannot be 0.

Taking the first step

The rational root theorem creates a list of numbers that may be roots of a particular polynomial. After using the theorem to make your list of potential roots, you plug the numbers into the polynomial to determine which, if any, work. You may run across an instance where none of the candidates work, which tells you that there are no rational roots. (And if a given rational number isn't on the list of possibilities that you come up with, it can't be a root of that polynomial.)

Before you start to plug and chug, however, check out the "Putting Descartes in charge of signs" section, later in this chapter — it helps you with your guesses. Also, you can refer to "Finding Roots Synthetically," later in this chapter, for a quicker method than plugging in.

To find the rational roots of the polynomial $y = x^4 - 3x^3 + 2x^2 + 12$, for example, you test the following possibilities: ±1, ±2, ±3, ±4, ±6, and ±12. These values are all the factors of the number 12. Technically, you divide each of these factors of 12 by the factors of the lead coefficient, but because the lead coefficient is one (as in $1x^4$), dividing by that number won't change a thing.

Find the roots of the polynomial $y = 6x^7 - 4x^4 - 4x^3 + 2x - 20$.

You first list all the factors of 20: ±1, ±2, ±4, ±5, ±10, and ±20. Now divide each of those factors by the factors of 6. You don't need to bother dividing by 1 to create your list, but you need to divide each by 2, 3, and 6: $\pm\frac{1}{2}, \pm\frac{2}{2}, \pm\frac{4}{2}, \pm\frac{5}{2}, \pm\frac{10}{2}, \pm\frac{20}{2}, \pm\frac{1}{3}, \pm\frac{2}{3}, \pm\frac{4}{3}, \pm\frac{5}{3}, \pm\frac{10}{3}, \pm\frac{20}{3}, \pm\frac{1}{6}, \pm\frac{2}{6}, \pm\frac{4}{6}, \pm\frac{5}{6}, \pm\frac{10}{6}, \pm\frac{20}{6}$. And, of course, you include : ±1, ±2, ±4, ±5, ±10, and ±20 as candidates.

You may have noticed some repeats in the previous list that occur when you reduce fractions. You can discard the repeats. And, even though this looks like a mighty long list, between the integers and fractions, it still gives you a reasonable number of candidates to try out. You can check them off in a systematic manner.

Changing from roots to factors

When you have the factored form of a polynomial and set it equal to 0, you can solve for the solutions (or x-intercepts, if that's what you want). Just as important, if you have the solutions, you can go backward and write the factored form. Factored forms are needed when you have polynomials in the numerator and denominator of fractions and you want to reduce the fraction. Factored forms are easier to compare with one another.

How can you use the rational root theorem to factor a polynomial function? Why would you want to? The answer to the second question, first, is that you can reduce a factored form if it's in a fraction. Also, a factored form is more easily graphed. Now, for the first question: You use the rational root theorem to find roots of a polynomial and then translate those roots into binomial factors whose product is the polynomial.

If $x = \frac{b}{a}$ is a root of the polynomial $f(x)$, the corresponding binomial $(ax - b)$ is a factor.

Write the factorization of a polynomial with the five roots $x = 1$, $x = -2$, $x = 3$, $x = \frac{3}{2}$, and $x = -\frac{1}{2}$.

Applying the rule, you get $f(x) = (x - 1)(x + 2)(x - 3)(2x - 3)(2x + 1)$. Notice that the positive roots give factors of the form $x - c$, and the negative roots give factors of the form $x + c$, which comes from $x - (-c)$. This is just one polynomial with these five roots. You can write other polynomials by multiplying the factorization by some constant.

To show *multiple roots*, or roots that occur more than once, use exponents on the factors. For example, if the roots of a polynomial are $x = 0$, $x = 2$, $x = 2$, $x = -3$, $x = -3$, $x = -3$, $x = -3$, and $x = 4$, a corresponding polynomial is $f(x) = x(x - 2)^2(x + 3)^4(x - 4)$.

Putting Descartes in charge of signs

Descartes' rule of signs tells you how many positive and negative *real* roots you may find in a polynomial. A *real number* is just about any number you can think of. It can be positive or negative, rational or irrational. The only thing it can't be is imaginary.

Counting up the number of possible positive roots

The first part of the rule of signs helps you identify how many of the roots of a polynomial are positive.

Descartes' rule of signs (part I): The polynomial $f(x) = a_n x^n + a_{n-1}x^{n-1} + a_{n-2}x^{n-2} + \ldots + a_1 x^1 + a_0$ has at most n roots. Count the number of times the sign changes in the coefficients of f, and call that value p. The value of p is the maximum number of *positive* real roots of f. If the number of positive roots isn't p, it is $p - 2$, $p - 4$, or some number less by a multiple of 2.

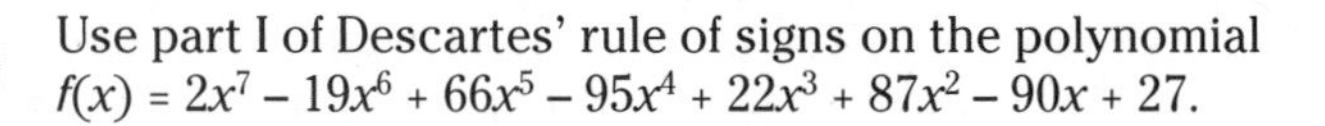

Use part I of Descartes' rule of signs on the polynomial $f(x) = 2x^7 - 19x^6 + 66x^5 - 95x^4 + 22x^3 + 87x^2 - 90x + 27$.

Count the number of sign changes. The sign of the first term starts as a positive, changes to a negative, and moves to positive; negative; positive; stays positive; negative; and then positive. Whew! In total, you count six sign changes. Therefore, you conclude that the polynomial has six positive roots, four positive roots, two positive roots, or none at all. When a root, such as $x = 3$ in the previous example, occurs more than once, you say that the root has *multiplicity* two or three or however many times it appears. This way, if you count the root as many times as it appears, the total will correspond to your predicted number.

Counting the possible number of negative roots

Along with the positive roots (see the previous section), Descartes' rule of signs deals with the possible number of negative roots of a polynomial. After you count the possible number of positive roots, you combine that value with the number of possible negative roots to make your guesses and solve the equation.

Descartes' rule of signs (part II): The polynomial $f(x) = a_n x^n + a_{n-1}x^{n-1} + a_{n-2}x^{n-2} + \ldots + a_1 x^1 + a_0$ has at most n roots. Find $f(-x)$, and then count the number of times the sign changes in $f(-x)$ and call that value q. The value of q is the maximum number of *negative* roots of f. If the number of negative roots isn't q, the number is $q - 2$, $q - 4$, and so on, for as many multiples of 2 as necessary. Again, you count a multiple root as many times as it occurs when applying the rule.

Determine the possible number of negative roots of the polynomial $f(x) = 2x^7 - 19x^6 + 66x^5 - 95x^4 + 22x^3 + 87x^2 - 90x + 27$.

You first find $f(-x)$ by replacing each x with $-x$ and simplifying:

$$f(-x) = 2(-x)^7 - 19(-x)^6 + 66(-x)^5 - 95(-x)^4 + 22(-x)^3 + 87(-x)^2 - 90(-x) + 27 = -2x^7 - 19x^6 - 66x^5 - 95x^4 - 22x^3 + 87x^2 + 90x + 27$$

As you can see, the function has only one sign change, from negative to positive. Therefore, the function has exactly one negative root — no more, no less. In fact, this negative root is -1.

Knowing the potential number of positive and negative roots for a polynomial is very helpful when you want to pinpoint an exact number of roots. The example polynomial I present in this section has only one negative real root. That fact tells you to concentrate your guesses on positive roots; the odds are better that you'll find a positive root first.

Finding Roots Synthetically

You use synthetic division to test the list of possible roots for a polynomial that you come up with by using the rational root theorem. *Synthetic division* is a method of dividing a polynomial by a binomial, using only the coefficients of the terms. The method is quick, neat, and highly accurate — usually even more accurate than long division, because it has fewer opportunities for "user error."

Using synthetic division when searching for roots

When you use synthetic division to look for roots in a polynomial, the last number on the bottom row of your synthetic division problem is the telling result. If that number is 0, the division had no remainder, and the number is a root. The fact that there's no remainder means that the binomial represented by the number is dividing the polynomial evenly. The number is a root because the binomial is a factor of the polynomial.

Use synthetic division, the rational root theorem, and Descartes' rule of signs to find roots of the polynomial $f(x) = x^5 + 5x^4 - 2x^3 - 28x^2 - 8x + 32$.

Using the rational root theorem, your list of the potential rational roots is ±1, ±2, ±4, ±8, ±16, and ±32.

Then, applying Descartes' rule of signs, you determine that there are two or zero positive real roots and three or one negative real roots.

Here are the steps for performing synthetic division on a polynomial to find its roots:

1. **Write the polynomial in order of decreasing powers of the exponents. Replace any missing powers with 0 to represent the coefficient.**

   In this case, you've lucked out. The polynomial is already in the correct order: $f(x) = x^5 + 5x^4 - 2x^3 - 28x^2 - 8x + 32$.

2. **Write the coefficients in a row, including the zeros.**

   1 5 –2 –28 –8 32

3. **Put the number you want to divide by in front of the row of coefficients, separated by a half-box. Then draw a horizontal line below the row of coefficients, leaving room for numbers under the coefficients.**

   In this case, my guess is $x = 1$.

   $$\underline{1}\rfloor \quad 1 \quad 5 \quad -2 \quad -28 \quad -8 \quad 32$$
   $$\overline{\qquad\qquad\qquad\qquad\qquad\qquad}$$

4. **Bring the first coefficient straight down below the line. Then multiply the number you bring below the line by the number that you're dividing into everything. Put the result under the second coefficient.**

1	1	5	–2	–28	–8	32
		1				
	1					

5. **Add the second coefficient and the product, putting the result below the line.**

1	1	5	–2	–28	–8	32
		1				
	1	6				

6. **Repeat the multiplication/addition with the rest of the coefficients.**

1	1	5	–2	–28	–8	32
		1	6	4	–24	–32
	1	6	4	–24	–32	0

The last entry on the bottom is a 0, so you know 1 is a root. Now, you can do a modified synthetic division when testing for the next root; you just use the numbers across the bottom. (These values are actually coefficients of the quotient, if you do long division; see the following section.)

If your next guess is to see if $x = -1$ is a root, the modified synthetic division appears as follows:

–1	1	6	4	–24	–32
		–1	–5	1	23
	1	5	–1	–23	–9

The last entry on the bottom row isn't 0, so –1 isn't a root.

The really good guessers amongst you decide to try $x = 2$, $x = -4$, $x = -2$, and $x = -2$ (a second time). These values represent the rest of the roots.

Synthetically dividing by a binomial

Finding the roots of a polynomial isn't the only excuse you need to use synthetic division. You can also use synthetic division to replace the long, drawn-out process of dividing a polynomial by a binomial. The polynomial can be any degree; the binomial has to be either $x + c$ or $x - c$, and the coefficient on the x is 1. This may seem rather restrictive, but a huge number of long divisions you'd have to perform fit in this category, so it helps to have a quick, efficient method to perform these basic division problems.

To use synthetic division to divide a polynomial by a binomial, you first write the polynomial in decreasing order of exponents, inserting a 0 for any missing exponent. The number you put in front or divide by is the *opposite* of the number in the binomial.

Divide $2x^5 + 3x^4 - 8x^2 - 5x + 2$ by the binomial $x + 2$ using synthetic division.

Using –2 in the synthetic division:

$$\begin{array}{r|rrrrrr} -2 & 2 & 3 & 0 & -8 & -5 & 2 \\ & & -4 & 2 & -4 & 24 & -38 \\ \hline & 2 & -1 & 2 & -12 & 19 & -36 \end{array}$$

As you can see, the last entry on the bottom row isn't 0. If you're looking for roots of a polynomial equation, this fact tells you that –2 isn't a root. In this case, because you're working on a long division application, the –36 is the remainder of the division — in other words, the division doesn't come out even.

You obtain the answer (quotient) of the division problem from the coefficients across the bottom of the synthetic division. You start with a power one value lower than the original polynomial's power, and you use all the coefficients, dropping the power by one with each successive coefficient. The last coefficient is the remainder, which you write over the divisor.

Here's the division problem and its solution. The original division problem is written first. Under the problem, you see the coefficients from the synthetic division written in front of variables — starting with one degree lower than the original problem. The remainder of –36 is written in a fraction on top of the divisor, $x + 2$.

$$\left(2x^5 + 3x^4 - 8x^2 - 5x + 2\right) \div \left(x + 2\right) =$$

$$2x^4 - x^3 + 2x^2 - 12x + 19 - \frac{36}{x + 2}$$

Chapter 8

Being Respectful of Rational Functions

In This Chapter

- Investigating domains and related vertical asymptotes
- Looking at limits and horizontal asymptotes
- Removing discontinuities of rational functions

The best way to investigate rational functions is to look at the intercepts, the asymptotes, any removable discontinuities, and the limits to tell where the function values have been, what they're doing for particular values of the domain, and what they'll be doing for large values of x. You also need all this information to discuss or graph a rational function.

Whether you're graphing rational functions by hand or with a graphing calculator, you need to be able to recognize the various characteristics (intercepts, asymptotes, and so on) of the rational function. And, if you don't know what these characteristics are and how to find them, your calculator is no better than a paperweight to you.

Examining Rational Functions

You see *rational functions* written, in general, in the form of a fraction:

$$y = \frac{f(x)}{g(x)}, \text{ where } f \text{ and } g \text{ are } \textit{polynomials}$$

Rational functions (and more specifically their graphs) are distinctive because of what they do and don't have. The graphs of rational functions *do* have *asymptotes* (dotted lines drawn in to help with the shape and direction of the curve), and the graphs often *don't* have all the real numbers in their domains. Polynomials and exponential functions (which I cover in Chapters 7 and 9, respectively) make use of all the real numbers — their domains aren't restricted.

Deliberating on domain

As I explain in Chapter 5, the *domain* of a function consists of all the real numbers that you can insert into the function equation. Values in the domain have to work in the equation and avoid producing imaginary or nonexistent answers.

The following list illustrates some examples of domains of rational functions:

- The domain of $y = \frac{x-1}{x-2}$ is all real numbers except 2.
- The domain of $y = \frac{x+1}{x(x+4)}$ is all real numbers except 0 and –4.

Investigating intercepts

Functions in algebra can have intercepts (where the graph of the function crosses or touches an axis). A rational function may have an x-intercept and/or a y-intercept, but it doesn't have to have either. You can determine whether a given rational function has intercepts by looking at its equation.

Introducing zero to find y-intercepts

The coordinates $(0, b)$ represent the y-intercept of a rational function. To find the value of b, you substitute a 0 for x and solve for y. If 0 is in the domain of a rational function, you can be sure that the function at least has a y-intercept.

Making X mark the spot

The coordinates $(a, 0)$ represent an x-intercept of a rational function. To find the value(s) of a, you let y equal 0 and solve for x. (Basically, you just set the numerator of the fraction equal to 0, after you completely reduce the fraction.)

Find the intercepts of the rational function $y = \frac{x^2 - 64}{x^2 + 3x - 4}$.

First, to find the *y*-intercept you replace each *x* with 0 to get $y = \frac{0 - 64}{0 + 0 - 4} = \frac{-64}{-4} = 16$. The *y*-intercept is (0, 16). To find the *x*-intercepts, set the numerator equal to 0 and solve for *x*. You get that $x^2 - 64 = (x - 8)(x + 8) = 0$. The two solutions are $x = 8$ and $x = -8$, so the *x*-intercepts are at (8, 0) and (–8, 0). When setting the numerator equal to 0 to get the *x*-intercepts, you need to be sure that none of the factors in the numerator is also in the denominator.

Assigning Roles to Asymptotes

The graphs of rational functions take on some distinctive shapes because of asymptotes. An asymptote is a sort of ghost line. *Asymptotes* are drawn into the graph of a rational function to show the shape and direction of the function's graph. The asymptotes aren't really part of the graphs. You lightly sketch in the asymptotes when you're graphing to help you with the final product. The types of asymptotes that you usually find in a rational function include the following:

- Vertical asymptotes
- Horizontal asymptotes
- Oblique (slant) asymptotes

In this section, I explain how you crunch the numbers of rational equations to identify asymptotes and graph them.

Validating vertical asymptotes

The equations of vertical asymptotes appear in the form $x = h$. This equation of a line has only the *x* variable — no *y* variable — and the number *h*. To find a vertical asymptote you establish, first, that in the rational function $y = \frac{f(x)}{g(x)}$, $f(x)$ and $g(x)$ have no common factors; then you determine when the denominator equals 0: $g(x) = 0$. The vertical asymptotes occur when the *x*-values make the denominator equal to 0.

Find the vertical asymptotes of the function $y = \frac{x}{x^2 - 4x + 3}$.

First note that there's no common factor in the numerator and denominator. Then set the denominator equal to 0. Factoring $x^2 - 4x + 3 = 0$, you get $(x - 1)(x - 3) = 0$. The solutions are $x = 1$ and $x = 3$, which are the equations of the vertical asymptotes.

Finding equations for horizontal asymptotes

The horizontal asymptote of a rational function has an equation that appears in the form $y = k$. This linear equation only has the variable y — no x — and the k is some number. A rational function has only one horizontal asymptote — if it has one at all (some rational functions have no horizontal asymptotes, others have one, and none of them has more than one). A rational function has a horizontal asymptote when the degree (highest power) of $f(x)$, the polynomial in the numerator, is less than or equal to the degree of $g(x)$, the polynomial in the denominator.

Here's a rule for determining the equation of a horizontal asymptote. The horizontal asymptote of $y = \frac{f(x)}{g(x)} = \frac{a_n x^n + a_{n-1}x^{n-1} + \cdots + a_0}{b_m x^m + b_{m-1}x^{m-1} + \cdots + b_0}$ is:

- $\boldsymbol{y = \frac{a_n}{b_m}}$ when $n = m$, meaning that the highest degrees of the polynomials are the same. The fraction here is made up of the lead coefficients of the two polynomials.
- $\boldsymbol{y = 0}$ when $n < m$, meaning that the degree in the numerator is less than the degree in the denominator.

Find the horizontal asymptote for $y = \frac{3x^4 - 2x^3 + 7}{x^4 - 3x^2 - 5}$.

The degree of the denominator is the same as the degree of the numerator. The horizontal asymptote is $y = 3$ (from $\frac{a_n}{b_m}$). The fraction formed by the lead coefficients is $y = \frac{3}{1} = 3$.

Taking vertical and horizontal asymptotes to graphs

When a rational function has one vertical asymptote and one horizontal asymptote, its graph usually looks like two flattened-out, C-shaped curves that appear diagonally opposite one another from the intersection of the asymptotes. Occasionally, the curves appear side by side, but that's the exception rather than the rule. Figure 8-1 shows you two examples of the more frequently found graphs in the one horizontal and one vertical classification.

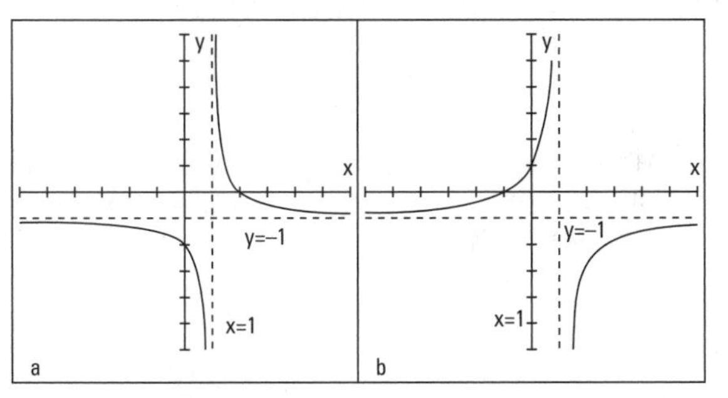

Figure 8-1: Rational functions approaching vertical and horizontal asymptotes.

Reconcile the rational equations $y = \frac{2-x}{x-1}$ and $y = \frac{x+1}{1-x}$ with the two graphs in Figure 8-1.

In both graphs, the vertical asymptotes are at $x = 1$, because the denominators are equal to 0 when $x = 1$. Also, in both graphs, the horizontal asymptotes are at $y = -1$.

In $y = \frac{2-x}{x-1}$, the highest power in both numerator and denominator is 1. You get $y = \frac{-1}{1}$. Letting $x = 0$, you get a y-intercept of (0, –2). Letting $y = 0$, you get an x-intercept of (2, 0). So this equation corresponds to Figure 8-1a.

The horizontal asymptote of function $y = \frac{x+1}{1-x}$ is $y = \frac{1}{-1}$. Letting $x = 0$, you get a y-intercept of (0, 1). Letting $y = 0$, you get an x-intercept of (–1, 0). So this equation corresponds to Figure 8-1b.

The graph of a rational function can cross a horizontal asymptote, but it never crosses a vertical asymptote. Horizontal asymptotes show what happens for very large or very small values of x.

Getting the scoop on oblique (slant) asymptotes

An *oblique* or *slant asymptote* takes the form $y = ax + b$. You may recognize this form as the slope-intercept form for the equation of a line. A rational function has a slant asymptote when the degree of the polynomial in the numerator is exactly one value greater than the degree in the denominator ($\frac{x^4}{x^3}$, for example).

You can find the equation of the slant asymptote by using long division. You divide the denominator of the rational function into the numerator and use the first two terms in the answer. Those two terms are the $ax + b$ part of the equation of the slant asymptote.

Find the slant asymptote of $y = \frac{x^4 - 3x^3 + 2x - 7}{x^3 + 3x - 1}$.

Doing the long division:

$$
\begin{array}{r}
x - 3 \qquad\qquad\qquad\quad \\
x^3 + 3x - 1 \overline{\big) x^4 - 3x^3 \qquad\quad + 2x - 7} \\
\underline{x^4 \qquad\quad + 3x^2 - x} \\
-3x^3 - 3x^2 + 3x - 7 \\
\underline{-3x^3 \qquad\quad - 9x + 3} \\
-3x^2 + 12x - 10
\end{array}
$$

You can ignore the remainder at the bottom. The slant asymptote for this example is $y = x - 3$. (For more on long division of polynomials, see *Algebra Workbook For Dummies,* by yours truly and published by Wiley.)

An oblique (or slant) asymptote creates two new possibilities for the graph of a rational function. If a function has an oblique asymptote, its curve tends to be a very-flat C on opposite sides of the intersection of the slant asymptote and a vertical asymptote (see Figure 8-2a), or the curve has U-shapes between the asymptotes (see Figure 8-2b).

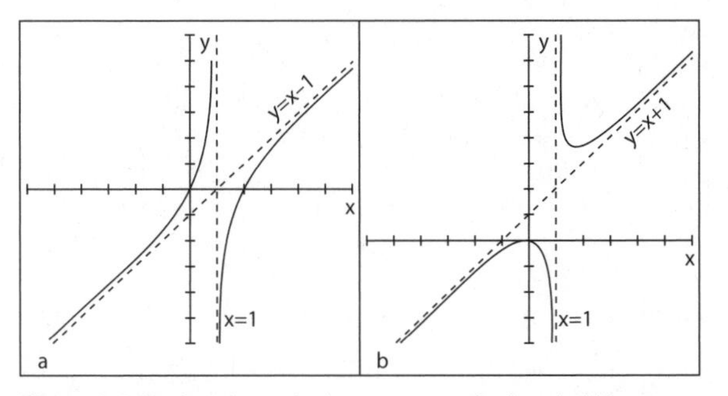

Figure 8-2: Rational graphs between vertical and oblique asymptotes.

Figure 8-2a has a vertical asymptote at $x = 1$ and a slant asymptote at $y = x - 1$; its intercepts are at (0, 0) and (2, 0). Figure 8-2b has a vertical asymptote at $x = 1$ and a slant asymptote at $y = x + 1$; its only intercept is at (0, 0).

Discounting Removable Discontinuities

A *discontinuity* in the graph of a function is just what the word suggests: a break or pause in the action. Vertical asymptotes mark discontinuities. The domain does not include values at the vertical asymptotes. But rational functions can sometimes have *removable discontinuities.* The removable designation is, however, a bit misleading. The gap in the domain still exists at that "removable" spot, but the function values and graph of the curve tend to behave a little better at *x*-values where there's a removable discontinuity. The function values stay close together — they don't spread far apart — and the graphs just have tiny holes, not vertical asymptotes where the graphs rise to positive infinity or plunge to negative infinity.

Removable discontinuities are found when you're factoring the original function statement — if it does factor. If the numerator and denominator don't have a common factor, then there isn't a removable discontinuity.

Finding removable discontinuities by factoring

Discontinuities are *removed* when they no longer have an effect on the rational function equation. You know this is the case when you find a factor that's common to both the numerator and the denominator. You accomplish the removal process by factoring the polynomials in the numerator and denominator of the rational function and then reducing the fraction.

To remove the discontinuity in the rational function $y = \frac{x^2 - 4}{x^2 - 5x - 14}$, for example, you first factor the fraction and reduce:

$$y = \frac{(x-2)(x+2)}{(x-7)(x+2)} = \frac{(x-2)\cancel{(x+2)}}{(x-7)\cancel{(x+2)}}$$

The *removable discontinuity* occurs when $x = -2$. Now you have a new function statement:

$$y = \frac{x-2}{x-7}$$

By getting rid of the removable discontinuity, you simplify the equation that you're graphing. Now you need only work with the new equation which shows you a function with vertical asymptote of $x = 7$, horizontal asymptote of $y = 1$, y-intercept of $\left(0, \frac{2}{7}\right)$, and x-intercept of $(2, 0)$. You also will have a "hole" in the graph when $x = -2$. Substituting that x-value into the new equation, you get $y = \frac{4}{9}$. So the small hole in the graph, marking the discontinuity, is at $\left(-2, \frac{4}{9}\right)$.

Evaluating the removals

You need to take care when *removing* discontinuities. Numbers excluded from the domain stay excluded even after you remove the discontinuity. The function still isn't defined for any values you find before the procedure. It's just that the function behaves differently at the different types of discontinuities. When the graph of a function has a hole, the curve approaches the value, skips it, and goes on. It behaves in a reasonable fashion: The function values skip over the discontinuity, but the x-values can get really close to it. When a vertical asymptote appears, though, the discontinuity doesn't go away. The function values go haywire — they're unrestrained as the x-values get close.

Figure 8-3 shows a rational function with a vertical asymptote at $x = -2$ and a removable discontinuity at $x = 3$. The horizontal asymptote is the x-axis (written $y = 0$). Unfortunately, graphing calculators don't show the little hollow circles indicating removable discontinuities. Oh, sure, they leave a gap there, but the gap is only 1 pixel wide, so you can't see it with the naked eye. You just have to know that the discontinuity is there. We're still better than the calculators!

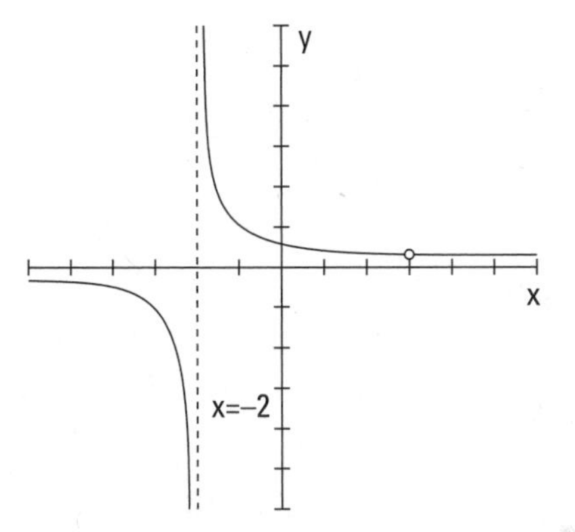

Figure 8-3: A removable discontinuity at the coordinates (3, 0.2).

Looking at Limits of Rational Functions

The *limit* of a rational function is a description of its behavior — telling you what the function equation is doing as you get nearer and nearer to some input value. If a function has a limit as you approach a number, then, as you get closer from the left or from the right, you home in on a particular function value. The function doesn't have to be defined at the number you're approaching (sometimes they are and sometimes not) — there could be a discontinuity at the point you're investigating. But, if a limit exists, the values of the function have a recognizable pattern.

Here is the special notation for limits:

$$\lim_{x \to a} f(x) = L$$

You read the notation as, "The limit of the function, $f(x)$, as x approaches the number a, is equal to L." The number a doesn't have to be in the domain of the function. You can talk about a limit of a function regardless of whether a is in the domain. And you can approach a; you don't actually reach it.

Determining limits at function discontinuities

The beauty of a limit is that it can also work when a rational function isn't defined at a particular number. The function $y = \frac{x-2}{x^2 - 2x}$, for example, is discontinuous at $x = 0$ and at $x = 2$. You find these numbers by factoring the denominator, setting it equal to 0 — $x(x - 2) = 0$ — and solving for x. This function has no limit when x approaches 0, but it has a limit when x approaches 2. Sometimes it's helpful to actually see the numbers — see what you get from evaluating a function at different values — so I've included Table 8-1. It shows what happens as x approaches 0 from the left and right, and it illustrates that the function has no limit at that value.

Table 8-1 Approaching $x = 0$ from Both Sides in $y = \frac{x-2}{x^2 - 2x}$

x Approaching 0 from the Left	*Corresponding Behavior of* $\lim_{x \to 0} \frac{x-2}{x^2 - 2x} \to -\infty$	*x Approaching 0 from the Right*	*Corresponding Behavior of* $\lim_{x \to 0} \frac{x-2}{x^2 - 2x} \to \infty$
–1.0	–1	1.0	1
–0.5	–2	0.5	2
–0.1	–10	0.1	10
–0.001	–1,000	0.001	1,000
–0.00001	–100,000	0.00001	100,000

Table 8-1 shows you that $\lim_{x \to 0} \frac{x-2}{x^2-2x}$ doesn't exist. As x approaches from below the value of 0, the values of the function drop down lower and lower toward negative infinity. Coming from above the value of 0, the values of the function raise higher and higher toward positive infinity. The sides will never come to an agreement; no limit exists.

Table 8-2 shows you how a function can have a limit even when the function isn't defined at a particular number. Sticking with the previous example function, you find a limit as x approaches 2.

Table 8-2 Approaching $x = 2$ from Both Sides in $y = \frac{x-2}{x^2-2x}$

x Approaching 2 from the Left	*Corresponding Behavior in* $\lim_{x \to 0} \frac{x-2}{x^2-2x} \to 0.5$	*x Approaching 2 from the Right*	*Corresponding Behavior in* $\lim_{x \to 0} \frac{x-2}{x^2-2x} \to 0.5$
1.0	1.0	3.0	0.3333 . . .
1.5	0.6666 . . .	2.5	0.4
1.9	0.526316 . . .	2.1	0.476190 . . .
1.99	0.502513 . . .	2.001	0.499750 . . .
1.999	0.500250 . . .	2.00001	0.4999975 . . .

Table 8-2 shows $\lim_{x \to 0} \frac{x-2}{x^2-2x} = 0.5$.

The numbers get closer and closer to 0.5 as x gets closer and closer to 2 from both directions. You find a limit at $x = 2$, even though the function isn't defined there.

Determining a limit algebraically

If you've examined the two tables from the previous section, you may think that the process of finding limits is exhausting. Allow me to tell you that algebra offers a much easier way to find limits — if they exist.

Functions with removable discontinuities have limits at the values where the discontinuities exist. In the "Finding removable discontinuities by factoring" section, earlier in this chapter, I show you the process needed to remove a discontinuity. I apply the same technique here.

To solve for the limit when $x = 2$ in the rational function $y = \frac{x-2}{x^2-2x}$, you first factor and then reduce the fraction:

$$y = \frac{x-2}{x^2-2x} = \frac{\cancel{x-2}}{x\cancel{(x-2)}} = \frac{1}{x}$$

Now you replace the x with 2 and get $y = 0.5$, the limit when $x = 2$. I write that as $\lim_{x \to 2} \frac{x-2}{x^2-2x} = 0.5$. In general, if a rational function factors, then you'll find a limit at the number excluded from the domain if the factoring makes that exclusion seem to disappear.

Determining whether the function has a limit

Some rational functions have limits at discontinuities and some don't. You can determine whether to look for a removable discontinuity in a particular function by first trying the x-value in the function. Replace all the x's in the function with the number in the limit (what x is approaching). The result of that substitution may tell you if you have a limit or not. You use the following rules of thumb:

- If $\lim_{x \to a} \frac{f(x)}{g(x)} = \frac{\text{some nonzero number}}{0}$, the function has no limit at a.
- If $\lim_{x \to a} \frac{f(x)}{g(x)} = \frac{0}{0}$, the function may have a limit at a. You reduce the fraction and evaluate the newly formed function equation at a.

Finding infinity

When a rational function doesn't have a limit at a particular value, the function values and graph have to go somewhere. Even though the function has no limit at some value, you can still say something about the behavior of the function. The behavior is described with *one-sided limits.*

A one-sided limit tells you what a function is doing as the *x*-value of the function approaches some number from one side or the other. One-sided limits are more restrictive; they only work from the left or from the right.

Here is the notation for indicating one-sided limits from the left or right:

- The limit as x approaches the value a from the left is $\lim\limits_{x \to a^-} f(x)$.
- The limit as x approaches the value a from the right is $\lim\limits_{x \to a^+} f(x)$.

Do you see the little positive or negative sign after the a? You can think of *from the left* as coming from the same direction as all the negative numbers on the number line and *from the right* as coming from the same direction as all the positive numbers.

Table 8-3 shows some values of the function $y = \frac{1}{x-3}$, which has a vertical asymptote at $x = 3$.

Table 8-3 Approaching $x = 3$ from Both Sides in $y = \frac{1}{x-3}$

Approaching 3 from the Left	*Corresponding Behavior in* $y = \frac{1}{x-3}$	*x Approaching 3 from the Right*	*Corresponding Behavior in* $y = \frac{1}{x-3}$
2.0	–1	4.0	1
2.5	–2	3.5	2
2.9	–10	3.1	10
2.999	–1,000	3.001	1,000
2.99999	–100,000	3.00001	100,000

You express the one-sided limits for the function from Table 8-3 as follows:

$$\lim_{x \to 3^-} \frac{1}{x-3} = -\infty, \quad \lim_{x \to 3^+} \frac{1}{x-3} = +\infty$$

The function goes down to negative infinity as it approaches 3 from below the value and up to positive infinity as it approaches 3 from above the value. "And nary the twain shall meet."

Looking at infinity

The previous section describes how function values can go to positive or negative infinity as x approaches some specific number. This section also talks about infinity, but it focuses on what rational functions do as their x-values become very large or very small (approaching infinity themselves).

A function such as the parabola $y = x^2 + 1$ opens upward. If you let x be some really big number, y gets very big, too. Also, when x is very small (a "big" negative number), you square the value, making it positive, so y is very big for the small x. In function notation, you describe what's happening to this function as the x-values approach infinity with $\lim_{x\to\infty}\left(x^2+1\right)=+\infty$.

In the case of rational functions, the limits at infinity — as x gets very large or very small — may be specific, finite, describable numbers. In fact, when a rational function has a horizontal asymptote, its limit at infinity is the same value as the number in the equation of the asymptote.

If you're looking for the horizontal asymptote of the function $y=\frac{4x^2+3}{2x^2-3x-7}$, for example, you can use the rules in the section "Taking vertical and horizontal asymptotes to graphs" to determine that the horizontal asymptote of the function is $y = 2$. Using limit notation, you can write the solution as $\lim_{x\to\infty}\frac{4x^2+3}{2x^2-3x-7}=2$.

The proper algebraic method for evaluating limits at infinity is to divide every term in the rational function by the highest power of x in the fraction and then look at each term. Here's an important property to use: As x approaches infinity, any term with $\frac{1}{x}$ or $\frac{1}{x^2}$ or $\frac{1}{x^3}$, and so on in it approaches 0 — in other words, gets very small — so you can replace those terms with 0 and simplify.

Here's how the property works when evaluating the limit of the previous example function, $y = \frac{4x^2+3}{2x^2-3x-7}$. The highest power of the variable in the fraction is x^2, so every term is divided by x^2:

$$\begin{aligned}\lim_{x\to\infty}\frac{4x^2+3}{2x^2-3x-7} &= \lim_{x\to\infty}\frac{\frac{4x^2}{x^2}+\frac{3}{x^2}}{\frac{2x^2}{x^2}-\frac{3x}{x^2}-\frac{7}{x^2}}\\ &= \lim_{x\to\infty}\frac{4+\frac{3}{x^2}}{2-\frac{3}{x}-\frac{7}{x^2}}\\ &= \frac{4+0}{2-0-0}\\ &= \frac{4}{2}\\ &= 2\end{aligned}$$

The limit as x approaches infinity is 2. As predicted, the number 2 is the number in the equation of the horizontal asymptote. The quick method for determining horizontal asymptotes is an easier way to find limits at infinity; this algebraic procedure is the correct *mathematical* way of doing it — and it shows why the other rule (the quick method) works.

Chapter 9

Examining Exponential and Logarithmic Functions

In This Chapter

- Investigating exponential functions and rules of exponents
- Introducing laws of logarithms and simplifications
- Solving exponential and logarithmic equations

Exponential growth and decay are natural phenomena. They happen all around us. And, being the thorough, worldly people they are, mathematicians have come up with ways of describing, formulating, and graphing these phenomena. You express the patterns observed when exponential growth and decay occur mathematically with exponential and logarithmic functions.

Computing Exponentially

An exponential function is unique because its variable appears in the exponential position and its constant appears in the base position. You write an *exponent,* or power, as a superscript just after the *base.* In the expression 3^x, for example, the variable x is the exponent, and the constant 3 is the base. The general form for an exponential function is $f(x) = a \cdot b^x$, where

- The base b is any positive number.
- The coefficient a is any real number (where $a \neq 0$).
- The exponent x is a variable representing any real number.

When you enter a number into an exponential function, you evaluate it by using the *order of operations,* evaluating the function in the following order:

1. **Powers and roots**
2. **Multiplication and division**
3. **Addition and subtraction**

Evaluate $f(x) = 4(3)^x + 1$ for $x = 2$ and $x = -2$.

Letting $x = 2$, you replace the x with the number 2. So, $f(2) = 4(3)^2 + 1 = 4(9) + 1 = 36 + 1 = 37$.

When $x = -2$, $f(-2) = 4(3)^{-2} + 1 = 4\left(\frac{1}{3^2}\right) + 1 = 4\left(\frac{1}{9}\right) + 1 = \frac{4}{9} + 1 = \frac{13}{9}$.

Getting to the Base of Exponential Functions

The base of an exponential function can be any positive number. The bigger the number, the bigger the function value becomes as the variable increases in value. (Sort of like the more money you have, the more money you make.) The bases can get downright small, too. In fact, when the base is some number between 0 and 1, you don't have a function that grows; instead, you have a function that falls.

Classifying bases

The base of an exponential function tells you so much about the nature and character of the function, making it one of the first things you should look for when working with exponential functions. One main distinguishing characteristic of bases of exponential functions is whether they're larger or smaller than 1. After you make that designation, you look at how *much* larger or how *much* smaller. The exponents also affect the expressions that contain them in somewhat predictable ways, making them another place to garner information about the function.

Because the domain of an exponential function is all real numbers, and the base is always positive, the result of b^x is always a positive number.

Focusing on bases

Algebra actually offers three classifications for the base of an exponential function, due to the fact that the numbers used as bases appear to react in distinctive ways when raised to positive powers:

- When $b > 1$, the values of b^x grow larger as x gets bigger — for example, $2^2 = 4$, $2^5 = 32$, $2^7 = 128$, and so on.
- When $b = 1$, the values of b^x show no movement. Raising the number 1 to higher powers always results in the number 1: $1^2 = 1$, $1^5 = 1$, $1^7 = 1$, and so on. You see no exponential growth or decay.
- When $0 < b < 1$, the value of b^x grows smaller as x gets bigger. Look at what happens to a fractional base when you raise it to the second, fifth, and eighth degrees: $\left(\frac{1}{3}\right)^2 = \frac{1}{9}$, $\left(\frac{1}{3}\right)^5 = \frac{1}{243}$, $\left(\frac{1}{3}\right)^8 = \frac{1}{6,561}$. The numbers get smaller and smaller as the powers get bigger.

Examining exponents

When an exponent is replaced with a particular type of real number, you get results that are somewhat predictable. The exponent makes the result take on different qualities, depending on whether the exponent is greater than 0, equal to 0, or smaller than 0:

- When the base $b > 1$ and the exponent $x > 0$, the values of b^x get bigger and bigger as x gets larger — for example, $4^3 = 64$ and $4^6 = 4,096$. You say that the values grow *exponentially.*
- When the base $b > 1$ and the exponent $x = 0$, the only value of b^x you get is 1. The rule is that $b^0 = 1$ for any number except $b = 0$. So, an exponent of 0 really flattens things out.
- When the base $b > 1$ and the exponent $x < 0$ — a negative number — the values of b^x get smaller and smaller as the exponents get further and further from 0. Take these expressions, for example: $6^{-1} = \frac{1}{6}$ and $6^{-4} = \frac{1}{6^4} = \frac{1}{1,296}$. These numbers can get very small very quickly.

Introducing the more frequently used bases: 10 and e

Exponential functions feature bases represented by numbers greater than 0. The two most frequently used bases are 10 and e, where $b = 10$ and $b = e$.

It isn't too hard to understand why mathematicians like to use base 10 — in fact, just hold all your fingers in front of your face! All the powers of 10 are made up of ones and zeros — for instance, $10^2 = 100$, $10^9 = 1{,}000{,}000{,}000$, and $10^{-5} = 0.00001$. How much more simple can it get? Our number system, the decimal system, is based on tens.

Like the value 10, base e occurs naturally. Members of the scientific world prefer base e because powers and multiples of e keep creeping up in models of natural occurrences. Including e's in computations also simplifies things for financial professionals, mathematicians, and engineers.

If you use a scientific calculator to get the value of e, you see only some of e. The numbers you see estimate only what e is; most calculators give you seven or eight decimal places such as these first nine decimal places: $e \approx 2.718281828$.

Exponential Equation Solutions

The process of solving exponential equations incorporates many of the same techniques you use in algebraic equations — adding to or subtracting from each side, multiplying or dividing each side by the same number, factoring, squaring both sides, and so on.

Solving exponential equations requires some additional techniques, however. One technique you use when solving exponential equations involves changing the original exponential equation into a new equation that has matching bases. Another technique involves putting the exponential equation into a more recognizable form — such as a linear or quadratic equation — and then using the appropriate methods.

Creating matching bases

If you see an equation written in the form $b^x = b^y$, where the same number represents the bases b, then it must be true that $x = y$. You read the rule as follows: "If b raised to the xth power is equal to b raised to the yth power, that implies that $x = y$."

Solve the equation $2^{3+x} = 2^{4x-9}$ for x.

You see that the bases (the twos) are the same, so the exponents must also be the same. You just solve the linear equation $3 + x = 4x - 9$ for the value of x: $12 = 3x$, or $x = 4$. You then put the 4 back into the original equation to check your answer: $2^{3+4} = 2^{4(4)-9}$, which simplifies to $2^7 = 2^7$, or $128 = 128$.

Many times, bases are related to one another by being powers of the same number.

Solve the equation $4^{x+3} = 8^{x-1}$ for x.

You need to write both the bases as powers of 2 and then apply the rules of exponents. The number 4 is equal to 2^2, and 8 is 2^3, so you can write the equation as: $\left(2^2\right)^{x+3} = \left(2^3\right)^{x-1}$.

Now, raising a power to a power gives you $2^{2x+6} = 2^{3x-3}$.

The bases are the same, so set the exponents equal to one another and solve for x: $2x + 6 = 3x - 3$, which solves to give you $x = 9$. Substituting the 9 for x in the original equation, you get

$$4^{9+3} = 8^{9-1}$$
$$4^{12} = 8^{8}$$
$$16{,}777{,}216 = 16{,}777{,}216$$

Quelling quadratic patterns

When exponential terms appear in equations with two or three terms, you may be able to treat the equations as you do quadratic equations (see Chapter 3) to solve them with familiar methods. Using the methods for solving quadratic equations is a big advantage because you can factor the exponential equations, or you can resort to the quadratic formula.

You can make use of just about any equation pattern that you see when solving the exponential functions. If you can simplify the exponential to the form of a quadratic or cubic and then factor, find perfect squares, find sums and difference of squares, and so on, you've made life easier by changing the equation into something recognizable and doable.

Factoring out a common factor

When you solve a quadratic equation by factoring out a greatest common factor (GCF), you use the rules of exponents to find the GCF and divide the terms.

Solve for x in $3^{2x} - 9 \cdot 3^x = 0$.

Factor 3^x from each term and get $3^x(3^x - 9) = 0$. Now use the *multiplication property of zero* (MPZ; see Chapter 1) by setting each of the separate factors equal to 0.

$3^x = 0$ has no solution; 3 raised to a power can't be equal to 0. But the second factor does not equal 0.

$$3^x - 9 = 0$$
$$3^x = 9$$
$$3^x = 3^2$$
$$x = 2$$

The factor is equal to 0 when $x = 2$; you find only one solution to the entire equation.

Factoring a quadratic-like trinomial

The trinomial $5^{2x} - 26 \cdot 5^x + 25 = 0$, resembles a quadratic trinomial that you can factor using unFOIL. This exponential equation has the same pattern as the quadratic equation $y^2 - 26y + 25 = 0$, which would look something like the exponential equation if you replace each 5^x with a y.

Solve for x in the equation $5^{2x} - 26 \cdot 5^x + 25 = 0$.

The quadratic $y^2 - 26y + 25 = 0$ factors into $(y - 1)(y - 25) = 0$. Using the same pattern on the exponential version, you get the factorization $(5^x - 1)(5^x - 25) = 0$. Setting each factor equal to 0, when $5^x - 1 = 0$, $5^x = 1$. This equation holds true when $x = 0$, making that one of the solutions. Now, when $5^x - 25 = 0$, you say that $5^x = 25$, or $5^x = 5^2$. In other words, $x = 2$. You find two solutions to this equation: $x = 0$ and $x = 2$.

Looking into Logarithmic Functions

A *logarithm* is actually the exponent of a number. Logarithmic (abbreviated *log*) functions are the inverses of exponential functions. Logarithms answer the question, "What power gave me that answer?" The log function associated with the exponential function $f(x) = 2^x$, for example, is $f^{-1}(x) = \log_2 x$. The superscript –1 after the function name f indicates that you're looking at the inverse of the function f. So, $\log_2 8$, for example, asks, "What power of 2 gave me 8?"

A logarithmic function has a *base* and an *argument*. The logarithmic function $f(x) = \log_b x$ has a base b and an argument x. The base must always be a positive number and not equal to 1. The argument must always be positive.

You can see how a function and its inverse work as exponential and log functions by evaluating the exponential function for a particular value and then seeing how you get that value back after applying the inverse function to the answer. For example, first let $x = 3$ in $f(x) = 2^x$; you get $f(3) = 2^3 = 8$. You put the answer, 8, into the inverse function $f^{-1}(x) = \log_2 x$, and you get $f^{-1}(8) = \log_2 8 = 3$. The answer comes from the definition of how logarithms work; the 2 raised to the power of 3 equals 8. You have the answer to the fundamental logarithmic question, "What power of 2 gave me 8?"

Presenting the properties of logarithms

Logarithmic functions share similar properties with their exponential counterparts. When necessary, the properties of logarithms allow you to manipulate log expressions so you can solve equations or simplify terms. As with exponential functions, the base b of a log function has to be positive. I show the properties of logarithms in Table 9-1.

Table 9-1 Properties of Logarithms

Property Name	*Property Rule*	*Example*
Equivalence	$y = \log_b x \leftrightarrow b^y = x$	$y = \log_9 3 \leftrightarrow 9^y = 3$
Log of a product	$\log_b xy = \log_b x + \log_b y$	$\log_2 8z = \log_2 8 + \log_2 z$
Log of a quotient	$\log_b \frac{x}{y} = \log_b x - \log_b y$	$\log_2 \frac{8}{5} = \log_2 8 - \log_2 5$
Log of a power	$\log_b x^n = n\log_b x$	$\log_3 8^{10} = 10\log_3 8$
Log of 1	$\log_b 1 = 0$	$\log_4 1 = 0$
Log of the base	$\log_b b = 1$	$\log_4 4 = 1$

Exponential terms that have a base e have special logarithms just for the e's (the ease?). Instead of writing the log base e as $\log_e x$, you insert a special symbol, ln, for the log. The symbol ln is called the *natural logarithm,* and it designates that the base is e. The equivalences for base e and the properties of natural logarithms are the same, but they look just a bit different. Table 9-2 shows them.

Table 9-2 Properties of Natural Logarithms

Property Name	*Property Rule*	*Example*
Equivalence	$y = \ln x \leftrightarrow e^y = x$	$6 = \ln x \leftrightarrow e^6 = x$
Natural log of a product	$\ln xy = \ln x + \ln y$	$\ln 4z = \ln 4 + \ln z$
Natural log of a quotient	$\ln \frac{x}{y} = \ln x - \ln y$	$\ln \frac{4}{z} = \ln 4 - \ln z$
Natural log of a power	$\ln x^n = n\ln x$	$\ln x^5 = 5\ln x$
Natural log of 1	$\ln 1 = 0$	$\ln 1 = 0$
Natural log of e	$\ln e = 1$	$\ln e = 1$

As you can see in Table 9-2, the natural logs are much easier to write — you have no subscripts. Professionals use natural logs extensively in mathematical, scientific, and engineering applications.

Doing more with logs than sawing

You can use the basic exponential/logarithmic equivalence $\log_b x = y \leftrightarrow b^y = x$ to simplify equations that involve logarithms. Applying the equivalence makes the equation much nicer. If you're asked to evaluate $\log_9 3$, for example (or if you have to change it into another form), you can write it as an equation, $\log_9 3 = x$, and use the equivalence: $9^x = 3$. Now you have it in a form that you can solve for x. (The x that you get is the answer or value of the original expression.)

Evaluate $\log_9 3$.

After writing $\log_9 3 = x$, and the equivalence $9^x = 3$, you solve by changing the 9 to a power of 3 and then finding x in the new, more familiar form:

$$\begin{aligned}(3^2)^x &= 3\\ 3^{2x} &= 3^1\\ 2x &= 1\\ x &= \frac{1}{2}\end{aligned}$$

The result tells you that $\log_9 3 = \frac{1}{2}$ — much simpler than the original log expression.

Evaluate $10(\log_3 27)$.

First, write $\log_3 27 = x$ and its equivalence, $3^x = 27$. The number 27 is 3^3, so you can say that $3^x = 3^3$. For that statement to be true, it must be that $x = 3$. Now, replacing $\log_3 27$ with 3 in the original problem, you get $10(\log_3 27) = 10(3) = 30$. Another way to approach evaluating $\log_3 27$ is to write it as $\log_3 3^3$. Using the law of logarithms involving powers (refer to Table 9-1), the expression becomes $3\log_3 3$. Again, using a law of logarithms from the same table, you can substitute 1 for $\log_3 3$, so $3\log_3 3 = 3(1) = 3$.

Using log laws to expand expressions

A big advantage of logs is their properties and the way that you can change powers, products, and quotients into simpler addition and subtraction. Put all the log properties together, and you can change a single complicated expression into several simpler terms.

Simplify $\log_3 \frac{x^3\sqrt{x^2+1}}{(x-2)^7}$ by using the properties of logarithms.

First, use the property for the log of a quotient and then use the property for the log of a product on the new first term.

$$\log_3 \frac{x^3\sqrt{x^2+1}}{(x-2)^7} = \log_3 x^3\sqrt{x^2+1} - \log_3(x-2)^7$$
$$= \log_3 x^3 + \log_3 \sqrt{x^2+1} - \log_3(x-2)^7$$

The last step is to use the log of a power on each term, changing the radical to a fractional exponent first:

$$\log_3 x^3 + \log_3(x^2+1)^{\frac{1}{2}} - \log_3(x-2)^7 =$$
$$3\log_3 x + \frac{1}{2}\log_3(x^2+1) - 7\log_3(x-2)$$

The three new terms you create are each much simpler than the whole expression.

Using compacting

Results of computations in science and mathematics can involve sums and differences of logarithms. When this happens, you usually prefer to have the answers written all in one term, which is where the properties of logarithms come in.

Simplify $4\ln(x+2) - 8\ln(x^2-7) - \frac{1}{2}\ln(x+1)$ by writing the three terms as a single logarithm.

First, apply the property involving the natural log (ln) of a power to all three terms. Then factor out –1 from the last two terms and write them in a bracket:

$$\ln(x+2)^4 - \ln(x^2-7)^8 - \ln(x+1)^{\frac{1}{2}} =$$
$$\ln(x+2)^4 - \left[\ln(x^2-7)^8 + \ln(x+1)^{\frac{1}{2}}\right]$$

Now use the property involving the ln of a product on the terms in the bracket, change the $\frac{1}{2}$ exponent to a radical, and use the property for the ln of a quotient to write everything as the ln of one big fraction:

$$\ln(x+2)^4 - \left[\ln(x^2-7)^8 + \ln(x+1)^{\frac{1}{2}}\right] =$$

$$\ln(x+2)^4 - \left[\ln(x^2-7)^8 (x+1)^{\frac{1}{2}}\right] =$$

$$\ln(x+2)^4 - \ln(x^2-7)^8 \sqrt{x+1} =$$

$$\ln\frac{(x+2)^4}{(x^2-7)^8 \sqrt{x+1}}$$

The expression is messy and complicated, but it sure is compact.

Solving Equations Containing Logs

Logarithmic equations can have one or more solutions, just like other types of algebraic equations. What makes solving log equations a bit different is that you get rid of the log part as quickly as possible, leaving you to solve either a polynomial or an exponential equation in its place. Polynomial and exponential equations are easier and more familiar, and you may already know how to solve them. The only caution I present before you begin solving logarithmic equations is that you need to check the answers you get from the new, revised forms. You may get answers to the polynomial or exponential equations, but they may not work in the logarithmic equation. Switching to another type of equation introduces the possibility of *extraneous roots* — answers that fit the new, revised equation that you choose but sometimes don't fit in with the original equation.

Seeing all logs created equal

One type of log equation features each term carrying a logarithm in it (and the logarithms have to have the same base). You can apply the following rule:

If $\log_b x = \log_b y$, then $x = y$.

Solve the equation $\log_4 x^2 = \log_4(x + 6)$.

Apply the rule so that you can write and solve the equation $x^2 = x + 6$. Setting the quadratic equation equal to 0, you get $x^2 - x - 6 = 0$ which factors into $(x - 3)(x + 2) = 0$. The solutions $x = 3$ and $x = -2$ are for the quadratic equation, and both work in the original logarithmic equation. You always must check, though, because the solutions from a related quadratic equation don't always work in the original.

The following equation shows you how you may get an extraneous solution. Note that, when there's no base showing, you assume that you have common logarithms that are base 10.

Solve $\log(x - 8) + \log(x) = \log(9)$.

First apply the property involving the log of a product to get just one log term on the left: $\log(x - 8)(x) = \log(9)$. Next, you use the property that allows you to drop the logs and get the equation $(x - 8)x = 9$. This is a quadratic equation that you can solve with factoring. Multiplying on the left, you get $x^2 - 8x$. Subtracting 9 from each side, the quadratic equation is $x^2 - 8x - 9 = 0$, which factors into $(x - 9)(x + 1) = 0$. The two solutions of the quadratic equation are $x = 9$ and $x = -1$.

Checking the answers, you find that the solution 9 works just fine, but the -1 doesn't work: $\log(-1 - 8) + \log(-1) = \log(9)$. You can stop right there. Both of the logs on the left have negative arguments. The argument in a logarithm has to be positive, so the -1 doesn't work in the log equation (even though it was just fine in the quadratic equation). You determine that -1 is an extraneous solution and throw it out.

Solving log equations by changing to exponentials

When a log equation has log terms and a term that doesn't have a logarithm in it, you need to use algebra techniques and log properties (see Table 9-1) to put the equation in the form $y = \log_b x$. After you create the right form, you can apply the equivalence to change it to a purely exponential equation.

Solve $\log_3(x + 8) - 2 = \log_3 x$.

First subtract $\log_3 x$ from each side and add 2 to each side to get $\log_3(x + 8) - \log_3 x = 2$. Now you apply the property involving the log of a quotient, rewrite the equation by using the equivalence, and solve for x:

$$\log_3 \frac{x+8}{x} = 2$$

$$3^2 = \frac{x+8}{x}$$

$$9x = x + 8$$

$$8x = 8$$

$$x = 1$$

The only solution is $x = 1$, which works in the original logarithmic equation.

Chapter 10

Getting Creative with Conics

In This Chapter

- Determining the centers of circles, ellipses, and hyperbolas
- Graphing parabolas using vertex and direction
- Using equations to sketch graphs of all conics

Conic is the name given to a special group of curves. The four conic sections are a parabola, circle, ellipse, and hyperbola.

Each conic section has a specific form or type of equation, and I cover each in this chapter. You can glean a good deal of valuable information from a conic section's equation, such as where it's centered in a graph, how wide it opens, and its general shape. I also discuss the techniques that work best for you when you're called on to graph conics.

The graphs of circles and ellipses are closed curves. Parabolas and hyperbolas open upward, downward, left, or right — depending on the type you're graphing. Just to acquaint you with what conic sections look like, I show you some graphs in Figure 10-1. Then, in subsequent sections, I give you all the details in terms of the characteristics and important features of the individual conics.

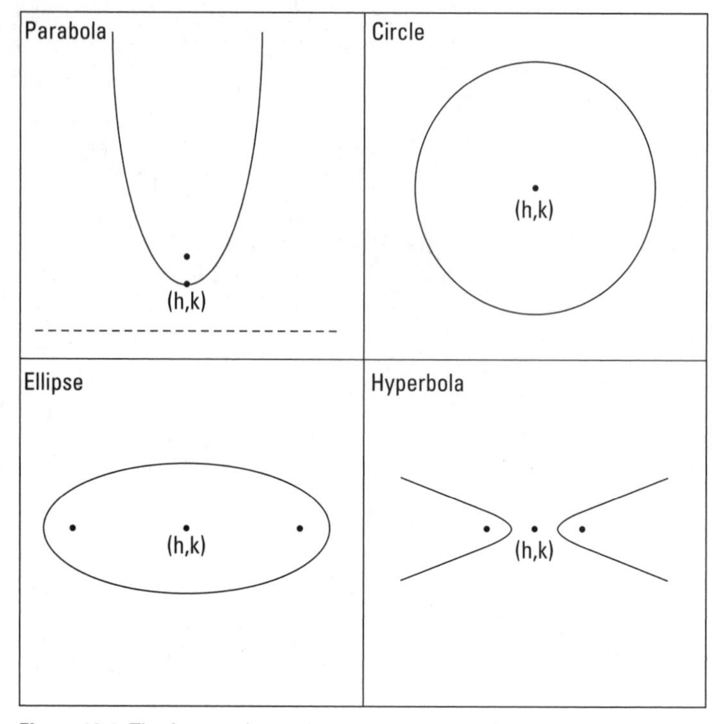

Figure 10-1: The four conic sections.

Posing with Parabolas

A *parabola,* a U-shaped conic that I first introduce in Chapter 6 (the parabola is the only conic section that fits the definition of a polynomial), is defined as all the points that fall the same distance from some fixed point, called its *focus,* and a fixed line, called its *directrix.* The focus is denoted by F, and the directrix by $y = d$ (assuming the parabola opens up or down).

A parabola has a couple other defining features. The *axis of symmetry* of a parabola is a line that runs through the focus and is perpendicular to the directrix. The axis of symmetry does just what its name suggests: It shows off how symmetric a parabola is. A parabola is a mirror image on either side of its axis. Another feature is the parabola's *vertex.* The vertex is the curve's extreme point — the lowest or highest point, or the point on the curve farthest right or farthest left. The vertex is also the point where the axis of symmetry crosses the curve.

Generalizing the form of a parabola's equation

The curves of parabolas can open upward, downward, to the left, or to the right; they also can be steep (tight) or widespread. The vertex can be anywhere in the coordinate plane. So, how do you track the curves down to pin them on a graph? You look to their equations, which give you all the information you need to find out where they've wandered to.

Opening left or right

When the vertex of a parabola is at the point (h, k), and the general form for the equation is as follows, the parabola opens left or right:

$$(y - k)^2 = 4a(x - h)$$

When the y variable is squared, the parabola opens left or right. From this equation, you can extract information about the elements:

- If $4a$ is positive, the curve opens right; if $4a$ is negative, the curve opens left.
- If $|4a| > 1$, the parabola is relatively wide; if $|4a| < 1$, the parabola is relatively narrow.
- The focus is at the point $(h + a, k)$.
- The directrix is $x = h - a$.

Opening up or down

When the vertex of a parabola is at the point (h, k), and the general form for the equation is as follows, the parabola opens up or down:

$$(x - h)^2 = 4a(y - k)$$

When the x variable is squared, the parabola opens up or down. Here's the info you can extract from this equation:

- If $4a$ is positive, the parabola opens upward; if $4a$ is negative, the curve opens downward.
- If $|4a| > 1$, the parabola is wide; if $|4a| < 1$, the parabola is narrow.
- The focus is at the point $(h, k + a)$.
- The directrix is $y = k - a$.

Making short work of a parabola's sketch

Parabolas have distinctive U-shaped graphs, and with just a little information, you can make a relatively accurate sketch of the graph of a particular parabola. The first step is to think of all parabolas as being in one of the general forms I list in the previous section.

Here's the full list of steps to follow when sketching the graph of a parabola — either $(x - h)^2 = 4a(y - k)$ or $(y - k)^2 = 4a(x - h)$:

1. **Determine the coordinates of the vertex (h, k) and plot that vertex.**

   If the equation contains $(x + h)$ or $(y + k)$, change the forms to $(x - [-h])$ or $(y - [-k])$, respectively, to determine the correct signs. Actually, you're just reversing the sign that's already there.

2. **Determine the direction the parabola opens, and decide if it's wide or narrow, by looking at the $4a$ portion of the general parabola equation.**

3. **Lightly sketch in the axis of symmetry that goes through the vertex.**

   $x = h$ when the parabola opens up or down and $y = k$ when it opens left or right.

4. **Choose a couple other points on the parabola and find each of their partners on the other side of the axis of symmetry to help you with the sketch.**

For example, if you want to graph the parabola $(y + 2)^2 = 8(x - 1)$, you first note that this parabola has its vertex at the point (1, –2) and opens to the right, because the y is squared (if the x had been squared, it would open up or down) and a, being 2, is positive. The graph is relatively wide about the axis of symmetry, $y = -2$, because $a = 2$, which is greater than 1. Figure 10-2a shows the vertex, axis of symmetry, and two points that satisfy the equation of the parabola. You find the points by substituting in a value for y and solving for x.

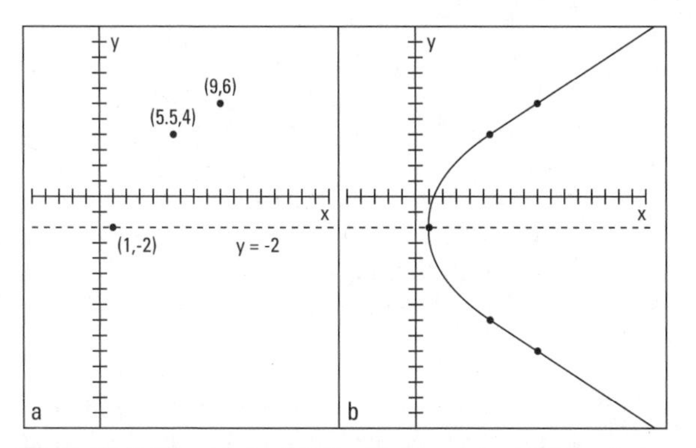

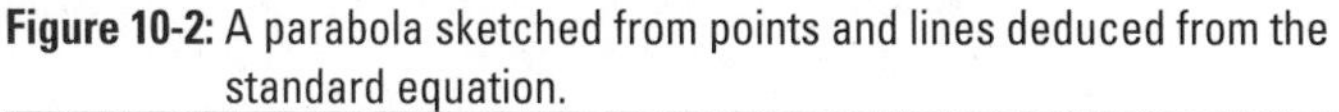

Figure 10-2: A parabola sketched from points and lines deduced from the standard equation.

The two randomly chosen points have counterparts on the opposite side of the axis of symmetry. The point (9, 6) is 8 units above the axis of symmetry, so 8 units below the axis puts you at (9, –10). The point (5.5, 4) is 6 units above the axis of symmetry, so its partner is the point (5.5, –8). Figure 10-2b shows the two new points and the parabola sketched in.

Changing a parabola's equation to the standard form

When the equation of a parabola appears in standard form, you have all the information you need to graph it or to determine some of its characteristics, such as direction or size. Not all equations come packaged that way, though. You may have to do some work on the equation first to be able to identify anything about the parabola.

The standard form of a parabola is $(x - h)^2 = 4a(y - k)$ or $(y - k)^2 = 4a(x - h)$, where (h, k) is the vertex.

The methods used here to rewrite the equation of a parabola into its standard form also apply when rewriting equations of circles, ellipses, and hyperbolas. The standard forms for conic sections are factored forms that allow you to immediately identify needed information. Different algebra situations call for different standard forms — the form just depends on what you need from the equation.

For example, if you want to convert the equation $x^2 + 10x - 2y + 23 = 0$ into the standard form, you act out the following steps, which contain a method called *completing the square,* which I show you here.

1. **Rewrite the equation with the x^2 and x terms (or the y^2 and y terms) on one side of the equation and the rest of the terms on the other side.**

   $x^2 + 10x = 2y - 23$

2. **Add a number to each side to make the side with the squared term into a perfect square trinomial (thus, completing the square).**

   $x^2 + 10x + 25 = 2y - 23 + 25$

3. **Rewrite the perfect square trinomial in factored form, and factor the terms on the other side by the coefficient of the variable.**

   $(x + 5)^2 = 2y + 2$

   $(x + 5)^2 = 2(y + 1)$

You now have the equation in standard form. The vertex is at $(-5, -1)$; it opens upward and is fairly wide.

Circling Around a Conic

A *circle,* probably the most recognizable of the conic sections, is defined as all the points plotted at the same distance from a fixed point — the circle's center, (h, k). The fixed distance is the radius, r, of the circle.

The standard form for the equation of a circle with radius r and with its center at the point (h, k) is $(x - h)^2 + (y - k)^2 = r^2$.

When the equation of a circle appears in the standard form, it provides you with all you need to know about the circle: its center and radius. With these two bits of information, you can sketch the graph of the circle. The equation $x^2 + y^2 + 6x - 4y - 3 = 0$, for example, is the equation of a circle. You can change this equation to the standard form by *completing the square* for each of the variables. Just follow these steps:

1. **Change the order of the terms so that the x's and y's are grouped together and the constant appears on the other side of the equal sign.**

   Leave a space after the groupings for the numbers that you need to add:

   $x^2 + 6x \quad + y^2 - 4y \quad = 3$

2. **Complete the square for each variable, adding the numbers that create perfect square trinomials.**

   $x^2 + 6x + 9 + y^2 - 4y + 4 = 3 + 9 + 4$

3. **Factor each perfect square trinomial.**

   $(x + 3)^2 + (y - 2)^2 = 16$

The example circle has its center at the point (–3, 2) and has a radius of 4 (the square root of 16). To sketch this circle, you locate the point (–3, 2) and then count 4 units up, down, left, and right; sketch in a circle that includes those points.

Getting Eclipsed by Ellipses

The ellipse is considered the most aesthetically pleasing of all the conic sections. It has a nice oval shape often used for mirrors, windows, and art forms.

The definition of an *ellipse* is all the points where the sum of the distances from the points to two fixed points is a constant. The two fixed points are the *foci* (plural of *focus*), denoted by *F*. Figure 10-3 illustrates this definition. You can pick a point on the ellipse, and the two distances from that point to the two foci add up to the same number as the sum of the distances from any other point on the ellipse to the foci. In Figure 10-3, the distances from point *A* to the two foci are 3.2 and 6.8, which add up to 10. The distances from point *B* to the two foci are 5 and 5, which also add up to 10.

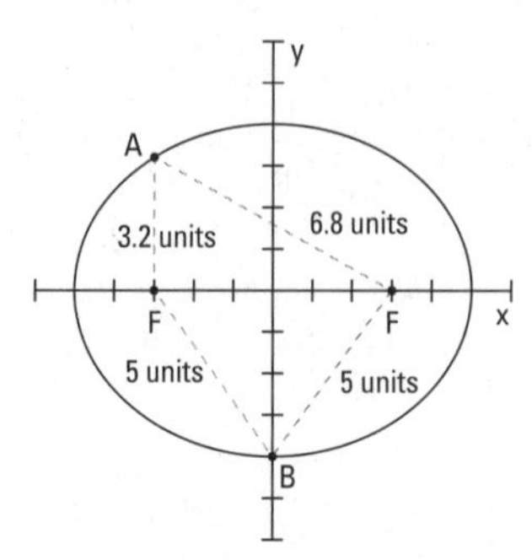

Figure 10-3: The sum of the two distances to the foci are the same.

You can think of the ellipse as a sort of squished circle. Of course, there's much more to ellipses than that, but the label sticks because the standard equation of an ellipse has a vague resemblance to the equation for a circle (see the previous section).

The standard equation for an ellipse with its center at the point (*h*, *k*) is $\frac{(x-h)^2}{a^2}+\frac{(y-k)^2}{b^2}=1$, where

- (x, y) is a point on the ellipse.
- a is half the length of the ellipse from left to right at its widest point.
- b is half the distance up or down the ellipse at its tallest point.

The standard equation tells you about the center, whether the ellipse is long and narrow or tall and slim. The equation tells you how long across, and how far up and down. You may even want to know the coordinates of the foci. You can determine all these elements from the equation.

Determining the shape

An ellipse is crisscrossed by a *major axis* and a *minor axis*. Each axis divides the ellipse into two equal halves, with the *major axis* being the longer of the segments. The two axes intersect at the center of the ellipse. At the ends of the major axis, you find the *vertices* of the ellipse. Figure 10-4 shows two ellipses with their axes and vertices identified.

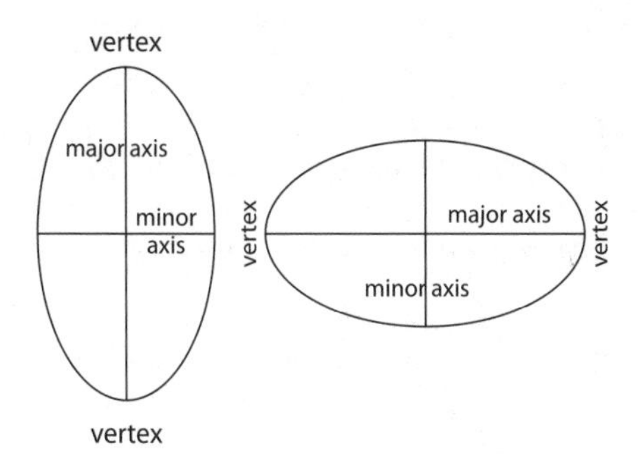

Figure 10-4: Ellipses with their axis properties identified.

To determine the shape of an ellipse, you need to pinpoint two characteristics:

- **Lengths of the axes:** You can determine the lengths of the two axes from the standard equation of the ellipse. You take the square root of the numbers in the denominators of the fractions. Whichever value is larger, a^2 or b^2, tells you which one is the major axis. The square roots of these numbers represent the distances from the center to the points on the ellipse along their respective axes. In other words, a is half the length of one axis, and b is half the length of the other. Therefore, $2a$ and $2b$ are the lengths of the axes.
- **Assignment of the axes:** The positioning of the axes is significant. The denominator that falls under the x's signifies the axis that runs parallel to the x-axis. The denominator that falls under the y factor signifies the axis that runs parallel to the y-axis.

Finding the foci

You can find the two foci of an ellipse by using information from the standard equation. The foci, for starters, always lie on the major axis. They lie c units from the center. To find the value of c, you use parts of the ellipse equation to form the equation $c^2 = a^2 - b^2$ or $c^2 = b^2 - a^2$, depending on which is larger, a^2 or b^2. The value of c^2 has to be positive.

In the ellipse $\frac{x^2}{25} + \frac{y^2}{9} = 1$, for example, the major axis runs across the ellipse, parallel to the x-axis. Actually, the major axis *is* the x-axis, because the center of this ellipse is the origin. You know this because the h and k are missing from the equation (actually, they're both equal to 0). You find the foci of this ellipse by solving the foci equation:

$$\begin{aligned} c^2 &= a^2 - b^2 \\ c^2 &= 25 - 9 \\ c^2 &= 16 \\ c &= \pm\sqrt{16} \\ &= \pm 4 \end{aligned}$$

So, the foci are 4 units on either side of the center of the ellipse. In this case, the coordinates of the foci are (–4, 0) and (4, 0).

Getting Hyped for Hyperbolas

The hyperbola is a conic section that features two completely disjoint curves, or *branches,* that face away from one another but are mirror images across a line that runs halfway between them.

A *hyperbola* is defined as all the points such that the difference of the distances from the point to two fixed points (called *foci*) is a positive constant value. In other words, you pick a value, such as the number 6; you find two distances whose difference is 6, such as 10 and 4; and then you find a point that rests 10 units from the one point and 4 units from the other point. The hyperbola has two axes, just as the ellipse has two axes (see the previous section). The axis of the hyperbola that goes through its two foci is called the *transverse axis.* The other axis, the *conjugate axis,* is perpendicular to the transverse axis, goes through the center of the hyperbola, and acts as the mirror line for the two branches.

There are two basic equations for hyperbolas. You use one when the hyperbola opens to the left and right: $\frac{(x-h)^2}{a^2} - \frac{(y-k)^2}{b^2} = 1$. You use the other when the hyperbola opens up and down: $\frac{(y-k)^2}{a^2} - \frac{(x-h)^2}{b^2} = 1$.

In both cases, the center of the hyperbola is at (h, k), and the foci are c units away from the center, where the relationship $b^2 = c^2 - a^2$ describes the relationship between the different parts of the equation.

Including the asymptotes

A very helpful tool you can use to sketch hyperbolas is to first lightly sketch in the two diagonal asymptotes of the hyperbola. *Asymptotes* aren't actual parts of the graph; they just help you determine the shape and direction of the curves. The asymptotes of a hyperbola intersect at the center of the hyperbola. You find the equations of the asymptotes by replacing the 1 in the equation of the hyperbola with a 0 and simplifying the resulting equation into the equations of two lines.

Find the equations of the asymptotes of the hyperbola $\frac{(x-3)^2}{9} - \frac{(y+4)^2}{16} = 1$.

Change the 1 to 0, set the two fractions equal to one another, and take the square root of each side:

$$\frac{(x-3)^2}{9} - \frac{(y+4)^2}{16} = 0$$

$$\frac{(x-3)^2}{9} = \frac{(y+4)^2}{16}$$

$$\sqrt{\frac{(x-3)^2}{9}} = \pm\sqrt{\frac{(y+4)^2}{16}}$$

$$\frac{x-3}{3} = \pm\frac{y+4}{4}$$

Then you multiply each side by 12 to get the equations of the asymptotes in better form and consider the two cases — one using the positive sign, and the other using the negative sign. The equations of the two asymptotes that result are $y = \frac{4}{3}x - 8$ and $y = -\frac{4}{3}x$. Notice that the slopes of the lines are the opposites of one another.

Graphing hyperbolas

Hyperbolas are relatively easy to sketch, *if* you pick up the necessary information from the equations. To graph a hyperbola, use the following steps as guidelines:

1. **Determine if the hyperbola opens to the sides or up and down by noting whether the x term is first or second.**

   The x term first means it opens to the sides.

2. **Find the center of the hyperbola by looking at the values of h and k.**
3. **Lightly sketch in a rectangle twice as wide as the square root of the denominator under the x value and twice as high as the square root of the denominator under the y value.**

   The rectangle's center is the center of the hyperbola.

4. **Lightly sketch in the asymptotes through the vertices of the rectangle (see the preceding section).**
5. **Draw in the hyperbola, making sure it touches the midpoints of the sides of the rectangle.**

You can use these steps to graph the hyperbola $\frac{(x+2)^2}{9} - \frac{(y-3)^2}{16} = 1$. First, note that this hyperbola opens to the left and right because the x value comes first in the equation. The center of the hyperbola is at (–2, 3).

Now comes the mysterious rectangle. Starting at the center at (–2, 3), you count 3 units to the right and left of center (totaling 6), because twice the square root of 9 is 6. Now you count 4 units up and down from center, because twice the square root of 16 is 8. When the rectangle is in place, you draw in the asymptotes of the hyperbola, diagonally through the vertices (corners) of the rectangle. Lastly, with the asymptotes in place, you draw in the hyperbola, making sure it touches the sides of the rectangle at the midpoints and slowly gets closer and closer to the asymptotes as they get farther from the center. You can see the full hyperbola in Figure 10-5.

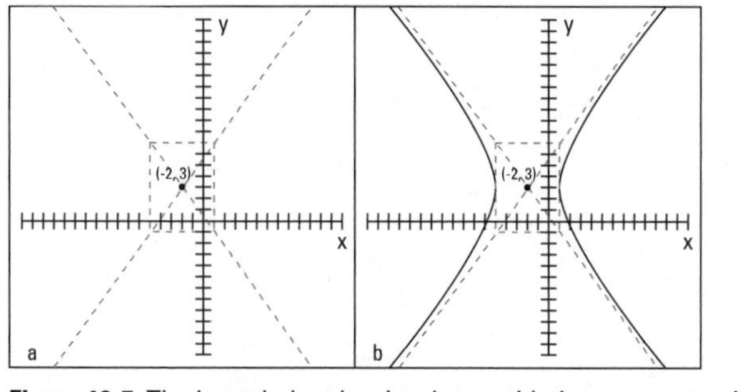

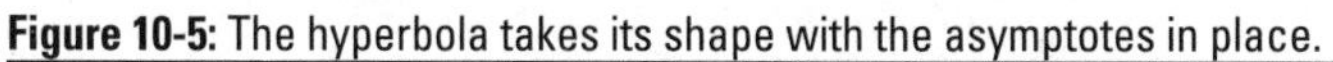

Figure 10-5: The hyperbola takes its shape with the asymptotes in place.

Chapter 11

Solving Systems of Equations

In This Chapter

- Finding solutions for systems of two, three, or more linear equations
- Determining if and where lines and parabolas intersect
- Expanding the search for intersections to other curves

A *system of equations* consists of a number of equations with common variables — variables that are linked in a specific way. The solution of a system of equations consists of the sets of numbers that make each equation in the system a true statement or a list of relationships between numbers that makes each equation in the system a true statement.

In this chapter, I cover both systems of linear equations and some nonlinear equations. You have a number of techniques at your disposal to solve systems of equations, including graphing lines, adding multiples of one equation to another, and substituting one equation into another.

Looking at Solutions Using the Standard Linear-Systems Form

The standard form for a system of linear equations is as follows:

$$\begin{cases} a_1x_1 + a_2x_2 + a_3x_3 + \ldots = k_1 \\ b_1x_1 + b_2x_2 + b_3x_3 + \ldots = k_2 \\ c_1x_1 + c_2x_2 + c_3x_3 + \ldots = k_3 \\ \vdots \end{cases}$$

The x's all represent variables, the k's are constants, and the a, b, c, and so on all represent constant coefficients of the variables.

If a system has only two linear equations with two variables, the equations appear in the $Ax + By = C$ form and can be graphed on the coordinate system to illustrate the solution. But a system of equations can contain any number of equations. (I show you how to work through larger systems in the later section, "Increasing the Number of Equations.")

Linear equations, like $Ax + By = C$, with two variables have lines as graphs. In order to solve a system of two linear equations with two variables, you need to determine what values for x and y make both the equations true at the same time. Your job is to account for which of three possible types of solutions (if you count "no solution" as a solution) can make this happen:

- **One solution:** The solution appears at the point where the lines intersect — the same x and the same y work at the same time in both equations.
- **An infinite number of solutions:** The equations are describing the same line.
- **No solution:** Occurs when the lines are parallel — no value for (x, y) works in both equations.

Solving Linear Systems by Graphing

To solve systems of linear equations with two equations and two variables (and integers as solutions), you can graph both equations on the same axes and you see one of three things: intersecting lines (one solution), identical lines (infinitely many solutions), or parallel lines (no solution).

Solving linear systems by graphing the lines created by the equations is very satisfying to your visual senses, but beware: Using this method to find a solution requires careful plotting of the lines. Also, the task of determining rational (fractions) or irrational (square roots) solutions from graphs on graph paper is too difficult, if not impossible. In general, solving systems by graphing isn't very practical.

Interpreting an intersection

Lines are made up of many, many points. When two lines cross one another, they share just one of those points. You need to graph very carefully, using a sharpened pencil and ruler with no bumps or holes.

A quick way to sketch two lines is to find their *intercepts* (where they cross the axes). Plot the intercepts on a graph and draw a line through them.

If the two lines clearly intersect at a point, you mark the point and determine the solution by counting the grid marks in the figure. This method shows you how important it is to graph the lines very carefully!

Tackling the same line

A unique situation that occurs with systems of linear equations happens when everything seems to work. Every point you find that works for one equation works for the other, too.

This match-made-in-heaven scenario plays out when the equations are just two different ways of describing the same line.

When two equations in a system of linear equations represent the same line, the equations are multiples of one another.

Putting up with parallel lines

Parallel lines never intersect and never have anything in common except their *slope.* So, when you solve systems of equations that have no solutions at all, you should know right away that the lines represented by the equations are parallel.

One way you can predict that two lines are parallel — and that no solution exists for the system of equations — is by checking the slopes of the lines. You can write each equation in *slope-intercept form* for the line. The slope-intercept form for the line $x + 2y = 8$, for example, is $y = -\frac{1}{2}x + 4$, and the slope-intercept form for $3x + 6y = 7$ is $y = -\frac{1}{2}x + \frac{7}{6}$. The lines both have the slope $-\frac{1}{2}$, and their *y*-intercepts are different, so you know the lines are parallel.

Using Elimination (Addition) to Solve Systems of Equations

Even though graphing lines to solve systems of equations is more visually satisfying, as a technique for solving systems of equations, graphing is time-consuming and requires careful plotting of points and cooperative answers. The two most preferred (and common) methods for solving systems of two linear equations are *elimination,* which I cover in this section, and *substitution,* which I cover in the section "Finding Substitution to Be a Satisfactory Substitute," later in this chapter. Determining which method you should use depends on what form the equations start out in and, often, personal preference.

To carry out the elimination method, you want to add two equations together, or subtract one from another other, and eliminate (get rid of) one of the variables. Sometimes you have to multiply one or both of the equations by a carefully selected number before you add them together (or subtract them).

Solve the following system of equations:

$$\begin{cases} 3x - 2y = 17 \\ 2x - 5y = 26 \end{cases}$$

The system requires some adjustments before you add or subtract the two equations. You have several different options to choose from to make the equations in this example system ready for elimination, and the one I would choose is to multiply the first equation by 2 and the second by –3 and then add to eliminate the x's.

Here's the new version of the system:

$$\begin{cases} 6x - 4y = 34 \\ -6x + 15y = -78 \end{cases}$$

Adding the two equations together, you get $11y = -44$, eliminating the x's. Dividing each side of the new equation by 11, you get $y = -4$. Substitute this value into the first *original* equation. Substituting –4 for the y value, you get $3x - 2(-4) = 17$. Solving for x, you get $x = 3$. Now check your work by putting the 3 and –4 into the second original equation. You get $2(3) - 5(-4) = 26$; $6 + 20 = 26$; $26 = 26$. Check! The solution is $(3, -4)$.

When you graph systems of two linear equations, it becomes pretty apparent when the systems produce parallel lines or have equations that represent the same line. But you can also recognize these situations algebraically, if you know what to look for.

- When doing the algebra using elimination or substitution and you get an *impossible statement,* such as $0 = 5$, then the false statement is your signal that the system doesn't have a solution and that the lines *are parallel.*

- If the algebra results in an equation that's *always* true, such as 0 = 0 or 5 = 5, then you know that the original equations are just two ways of giving you the *same line.*

Finding Substitution to Be a Satisfactory Substitute

Another method used to solve systems of linear equations is called *substitution.* Substitution works best when solving nonlinear systems, so some people prefer sticking to substitution for both types. The method used is often just a matter of personal choice.

Variable substituting made easy

Executing substitution in systems of two linear equations is a two-step process:

1. **Solve one of the equations for one of the variables, *x* or *y*.**
2. **Substitute the value of the variable into the other equation.**

Solve the following system by substitution:

$$\begin{cases} 2x - y = 1 \\ 3x - 2y = 8 \end{cases}$$

First look for a variable that is a likely candidate for the first step. In other words, you want to solve for it.

Look for a variable with a coefficient of 1 or –1, if possible. The y term of the first equation has a coefficient of –1, so you solve this equation for y (rewrite it so y is alone on one side of the equation). You get $y = 2x - 1$. Now you can substitute the $2x - 1$ for the y in the other equation:

$$\begin{aligned} 3x - 2y &= 8 \\ 3x - 2(2x - 1) &= 8 \\ 3x - 4x + 2 &= 8 \\ -x &= 6 \\ x &= -6 \end{aligned}$$

You've already created the equation $y = 2x - 1$, so you can put the value $x = -6$ into the equation to get y:

$$y = 2(-6) - 1 = -12 - 1 = -13$$

To check your work, put both values, $x = -6$ and $y = -13$, into the equation that you didn't change (the second equation, in this case): $3(-6) - 2(-13) = 8$; $-18 + 26 = 8$; $8 = 8$. Your work checks out. Your solution is $(-6, -13)$.

Writing solutions for coexisting lines

As I mention in the section "Recognizing situations with parallel and coexisting lines," earlier in this chapter, you want to identify the impossible (parallel lines) and always possible (coexisting lines). And then, with equations that represent the same line, you can say more about a solution.

The following system of equations represents two ways of saying the same equation — two equations that represent the same line:

$$\begin{cases} 3x - 2y = 4 \\ y = \dfrac{3}{2}x - 2 \end{cases}$$

When you solve the system by using substitution, you can end up with the equation: $4 = 4$.

When substitution creates an equation that's always true, any pair of values that works for one equation will work for the other. For this reason, you can write the solution in the (x, y) form, showing a pattern or formula for all the solutions.

In the following system, the y value is always 2 less than $\frac{3}{2}$ the x value (you get this from the second equation):

$$\begin{cases} 3x - 2y = 4 \\ y = \frac{3}{2}x - 2 \end{cases}$$

So the (x, y) form for the solution of the system is $\left(x, \frac{3}{2}x - 2\right)$.

Some solutions found, using the format, are: (2, 1), $\left(3, \frac{5}{2}\right)$, and (4, 4).

Taking on Systems of Three Linear Equations

Systems of three linear equations may also have solutions: sets of numbers (all the same for each equation) that make each of the equations true. What I show you in this section, involving three equations, can be extended to four, five, or even more equations. The basic processes are the same.

Finding the solution of a system of three linear equations

When you have a system of three linear equations and three unknown variables, you solve the system by reducing the three equations with three variables into a system of two equations with two variables. At that point, you're back to familiar territory and you have all sorts of methods at your disposal to solve the system (see the previous sections in this chapter). After you determine the values of the two variables in the new system, you *back-substitute* into one of the original equations to solve for the value of the third variable.

Solve the following system:

$$\begin{cases} 3x - 2y + z = 17 \\ 2x + y + 2z = 12 \\ 4x - 3y - 3z = 6 \end{cases}$$

First, you choose a variable to eliminate. The prime two candidates for elimination are the y and z because of the coefficients of 1 or –1 that occur in their equations. Assume that you choose to eliminate the z variable.

Start by multiplying the terms in the top equation by –2 and adding them to the terms in the middle equation. Then, multiply the terms in the top equation (the original top equation) by 3 and add them to the terms in the bottom equation:

$$\begin{aligned} -2(3x - 2y + z = 17) \rightarrow -6x + 4y - 2z &= -34 \\ \underline{2x + 4y + 2z} &\underline{= -12} \\ -4x + 5y \qquad &= -22 \end{aligned}$$

$$\begin{aligned} 3(3x - 2y + z = 17) \rightarrow 9x - 6y + 3z &= 51 \\ \underline{4x - 3y - 3z} &\underline{= 56} \\ 13x - 9y \qquad &= 57 \end{aligned}$$

Now deal with the two equations you created by solving them as a new system of equations with just two variables. Solve it by multiplying the terms in the first equation by 9 and the terms in the second equation by 5; add the two equations together, getting rid of the y terms, and solving for x:

$$\begin{aligned} -36x + 45y &= -198 \\ \underline{65x - 45y} &\underline{= \ \ 285} \\ 29x \qquad &= \ \ 87 \\ x &= \ \ 3 \end{aligned}$$

Now you substitute $x = 3$ into the equation $-4x + 5y = -22$. Choosing this equation is just an arbitrary choice — either equation will do. When you substitute $x = 3$, you get $-4(3) + 5y = -22$. Adding 12 to each side, you get $5y = -10$, or $y = -2$.

Putting $x = 3$ and $y = -2$ into the first equation, you get $3(3) - 2(-2) + z = 17$, giving you $9 + 4 + z = 17$. You subtract 13 from each side for a result of $z = 4$. Your solution is $x = 3$, $y = -2$, $z = 4$, or you can write it as an *ordered triple,* (3, –2, 4).

Generalizing with a system solution

When dealing with three linear equations and three variables, you may come across a situation where one of the equations is a linear combination of the other two. This means you won't find a single solution for the system — but you may find an infinite number of solutions or none at all. A generalized (giving infinitely many) solution looks like $(-z, 2z, z)$, where you can pick numbers for z that determine what the x and y values are.

You first get an inkling that a system has a generalized answer when you find out that one of the reduced equations you create is a multiple of the other.

Solve the following system:

$$\begin{cases} 2x + 3y - z = 12 \\ x - 3y + 4z = -12 \\ 5x - 6y + 11z = -24 \end{cases}$$

To solve this system, you can eliminate the z's by multiplying the terms in the first equation by 4 and adding them to the second equation. You then multiply the terms in the first equation by 11 and add them to the third equation:

$$\begin{array}{rl} 4(2x + 3y - z = 12) \rightarrow & 8x + 12y - 4z = 48 \\ & \underline{x - 3y + 4z = -12} \\ & 9x + 9y \quad\quad = 36 \end{array}$$

$$\begin{array}{rl} 11(2x + 3y - z = 12) \rightarrow & 22x + 33y - 11z = 132 \\ & \underline{5x - 6y + 11z = -24} \\ & 27x + 27y \quad\quad = 108 \end{array}$$

The second equation, $27x + 27y = 108$, is three times the first equation. Because these equations are multiples of one another, you know that the system has infinitely many solutions — not just a single solution.

To find those solutions, you take one of the equations and solve for a variable. You may choose to solve for y in $9x + 9y = 36$. Dividing through by 9, you get $x + y = 4$. Solving for y, you get $y = 4 - x$. You substitute that equation into one of the original equations in the system to solve for z in terms of x. After you solve for z this way, you have the three variables all written as some version of x.

Substituting $y = 4 - x$ into $2x + 3y - z = 12$, for example, you get

$$\begin{aligned} 2x + 3(4 - x) - z &= 12 \\ 2x + 12 - 3x - z &= 12 \\ -x - z &= 0 \\ z &= -x \end{aligned}$$

The ordered triple giving the solutions of the system is $(x, 4 - x, -x)$. You can find an infinite number of solutions, all determined by this pattern. Just pick an x, such as $x = 3$, and then the solution is $(3, 1, -3)$. These values of x, y, and z all work in the equations of the original system.

Increasing the Number of Equations

Systems of linear equations can be any size. You can have two, three, four, or even a hundred linear equations. (After you get past three or four, you definitely need to resort to technology.) Some of these systems have solutions and others don't. You have to dive in to determine whether you can find a solution or not. You can try to solve a system of just about any number of linear equations, but you find a single, unique solution (one set of numbers for the answer) only when the number of equations isn't smaller than the number of variables. If a system has three different variables, you need at least three different equations. Having enough equations for the variables doesn't guarantee a unique solution, but you have to at least start out that way.

The general process for solving n equations with n variables is to keep eliminating variables. A systematic way is to start with the first variable, eliminate it, move to the second variable, eliminate it, and so on until you create a reduced system with two equations and two variables. You solve for the solutions of that system and then start substituting values into the original equations. This process can be long and tedious, and errors are easy to come by, but if you have to do it by hand, this is a very effective method. Technology, however, is most helpful when systems get unmanageable.

The following system has five equations and five variables:

$$\begin{cases} x + y + z + w + t = 3 \\ 2x - y + z - w + 3t = 28 \\ 3x + y - 2z + w + t = -8 \\ x - 4y + z - w + 2t = 28 \\ 2x + 3y + z - w + t = 6 \end{cases}$$

You begin the process by eliminating the x's:

1. **Multiply the terms in the first equation by –2 and add them to the second equation.**
2. **Multiply the first equation through by –3 and add the terms to the third equation.**
3. **Multiply the first equation through by –1 and add the terms to the fourth equation.**
4. **Multiply the first equation through by –2 and add the terms to the last equation.**

After you finish (whew!), you get a system with the x's eliminated:

$$\begin{cases} -3y - z - 3w + t = 22 \\ -2y - 5z - 2w - 2t = -17 \\ -5y - 2w + t = 25 \\ y - z - 3w - t = 0 \end{cases}$$

Now you eliminate the y's in the new system by multiplying the last equation by 3, 2, and 5 and adding the results to the first, second, and third equations, respectively:

$$\begin{cases} -4z - 12w - 2t = 22 \\ -7z - 8w - 4t = -17 \\ -5z - 17w - 4t = 25 \end{cases}$$

You eliminate the z's in the latest system by multiplying the terms in the first equation by 7 and the second by –4 and adding them together. You then multiply the terms in the second equation by 5 and the third by –7 and add them together. The new system you create has only two variables and two equations:

$$\begin{cases} -52w + 2t = 222 \\ 79w + 8t = -260 \end{cases}$$

To solve the two-variable system in the most convenient way, you multiply the first equation through by –4 and add the terms to the second:

$$\begin{aligned} 208w - 8t &= -888 \\ \underline{79w + 8t} &\underline{= -260} \\ 287w \qquad &= -1148 \\ w &= -4 \end{aligned}$$

You find $w = -4$. Now back-substitute w into the equation $-52w + 2t = 222$ to get $-52(-4) + 2t = 222$, which simplifies to

$$\begin{aligned} 208 + 2t &= 222 \\ 2t &= 14 \\ t &= 7 \end{aligned}$$

Take these two values and plug them into $-4z - 12w - 2t = 22$. Substituting, you get $-4z - 12(-4) - 2(7) = 22$, which simplifies to

$$\begin{aligned} -4z + 34 &= 22 \\ -4z &= -12 \\ z &= 3 \end{aligned}$$

Put the three values into $y - z - 3w - t = 0$: $y - (3) - 3(-4) - 7 = 0$, or $y + 2 = 0$ and $y = -2$. Only one more to go!

Move back to the equation $x + y + z + w + t = 3$, and plug in values: $x + (-2) + 3 + (-4) + 7 = 3$, which simplifies to $x + 4 = 3$ and $x = -1$.

The solution reads: $x = -1$, $y = -2$, $z = 3$, $w = -4$, and $t = 7$ or, as an *ordered quintuple*, $(-1, -2, 3, -4, 7)$.

Intersecting Parabolas and Lines

A *parabola* is a predictable, smooth, U-shaped curve. A line is also very predictable; it goes up or down and left or right at the same rate forever and ever. If you put these two characteristics together, you can predict with a fair amount of accuracy what will happen when a line and a parabola share the same space.

When you combine the equations of a line and a parabola, you get one of three results:

- Two common solutions (intersecting in two places)
- One common solution (a line tangent to the parabola or parallel to the axis of symmetry)
- No solution at all (the line and parabola never cross)

The easiest way to find the common solutions, or common sets of values, for a line and a parabola is to solve their system of equations algebraically. A graph is helpful for confirming your work and putting the problem into perspective, but solving the system by graphing usually isn't very efficient. When solving a system of equations involving a line and a parabola, most mathematicians use the substitution method.

You almost always substitute x's for the y in an equation, because you often see functions written with the y's equal to so many x's. You may have to replace x's with y's, but that's the exception. Just be flexible.

Determining if and where lines and parabolas cross paths

The graphs of a line and a parabola can cross in two places, one place, or no place at all. In terms of equations, these assertions translate to two common solutions, one solution, or no solution at all. Doesn't that fit together nicely?

Taking on two solutions

A parabola and a line may have two points in common. When using the substitution method, you first need to solve for one or the other variable.

Find the intersections of $y = 3x^2 - 4x - 1$ and $x + y = 5$.

Solve for y in the equation of the line: $y = -x + 5$. Now you substitute this equivalence of y into the first equation, set the new equation equal to 0, and factor as you do any quadratic equation:

$$\begin{aligned} y &= 3x^2 - 4x - 1 \text{ and } y = -x + 5 \\ -x + 5 &= 3x^2 - 4x - 1 \\ 0 &= 3x^2 - 3x - 6 \\ 0 &= 3(x^2 - x - 2) \\ 0 &= 3(x - 2)(x + 1) \end{aligned}$$

Setting each of the binomial factors equal to 0, you get $x = 2$ and $x = -1$. When you substitute those values into the equation $y = -x + 5$, you find that when $x = 2$, $y = 3$, and when $x = -1$, $y = 6$. The two points of intersection, therefore, are (2, 3) and (–1, 6). Figure 11-1 shows the graphs of the parabola ($y = 3x^2 - 4x - 1$), the line ($y = -x + 5$), and the two points of intersection.

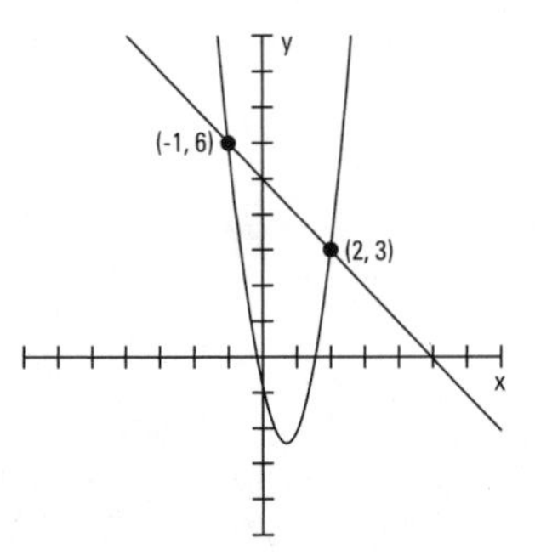

Figure 11-1: You find the two points of intersection with substitution.

Finding just one solution

When a line and a parabola have one point of intersection and, therefore, share one common solution, the line is *tangent* to the parabola or parallel to its axis of symmetry. A line and a curve can be tangent to one another if they touch or share exactly one point and if the line appears to follow the curvature at that point. (Two curves can also be tangent to one another — they touch at a point and then go their own merry ways.) The following example shows two figures that have only one point in common — at their point of tangency.

Find the intersection of $y = -x^2 + 5x + 6$ and $y = 3x + 7$.

Substitute the equivalence of y in the line equation into the parabolic equation and solve for x:

$$\begin{aligned} y &= -x^2 + 5x + 6 \text{ and } y = 3x + 7 \\ 3x + 7 &= -x^2 + 5x + 6 \\ 0 &= -x^2 + 2x - 1 \\ 0 &= -1(x^2 - 2x + 1) \\ 0 &= -1(x - 1)^2 \\ x &= 1 \end{aligned}$$

The dead giveaway that the parabola and line are tangent is the quadratic equation that results from the substitution. It has a *double root* — the same solution appears twice — when the binomial factor is squared.

Substituting $x = 1$ into the equation of the line, you get $y = 3(1) + 7 = 10$. The coordinates of the point of tangency are (1, 10).

Determining that there's no solution

You can see when no solution exists in a system of equations involving a parabola and line if you graph the two figures and find that their paths never cross. But you don't need to graph the figures to discover that a parabola and line don't intersect — the algebra gives you a "no-answer answer."

Solve the system of equations containing the parabola $x = y^2 - 4y + 3$ and the line $y = 2x + 5$.

Using substitution, you get the following:

$$x = y^2 - 4y + 3 \text{ and } y = 2x + 5$$
$$x = (2x + 5)^2 - 4(2x + 5) + 3$$
$$x = 4x^2 + 20x + 25 - 8x - 20 + 3$$
$$0 = 4x^2 + 11x + 8$$

The equation looks perfectly good so far, even though the quadratic doesn't factor. You have to resort to the quadratic formula. Substituting the numbers from the quadratic equation into the formula, you get the following:

$$\begin{aligned} x &= \frac{-11 \pm \sqrt{121 - 4(4)(8)}}{2(4)} \\ &= \frac{-11 \pm \sqrt{121 - 128}}{8} \\ &= \frac{-11 \pm \sqrt{-7}}{8} \end{aligned}$$

Whoa! You can stop right there. You see that a negative value sits under the radical. The square root of –7 isn't real, so no real-number answer exists for x. (For more on non-real numbers, see Chapter 12.) The nonexistent answer is your big clue that the system of equations doesn't have a common solution, meaning that the parabola and line never intersect.

I wish I could give you an easy way to tell that a system has no solution before you go to all that work. Think of it this way: An answer of *no solution* is a perfectly good answer.

Crossing Parabolas with Circles

The graph of a parabola is a U-shaped curve, and a circle — well, you could go 'round and 'round about a circle. When a parabola and circle share some of the same coordinate plane, they can interact in one of several different ways. The two figures can intersect at four different points, three points, two points, one point, or no points at all. The possibilities may seem endless, but that's wishful thinking. The five possibilities I list here are what you have to work with. Your challenge is to determine which situation you have and to find the solutions of the system of equations. And the best way to approach this problem is algebraically.

Finding multiple intersections

A parabola and a circle can intersect at up to four different points, meaning that their equations can have up to four common solutions. The next example shows you the algebraic solution of such a system of equations.

Find the four intersections of the parabola $y = -x^2 + 6x + 8$ and the circle $x^2 + y^2 - 6x - 8y = 0$.

To solve for the common solutions, you have to solve the system of equations by either substitution or elimination. You usually don't get to use elimination in problems like this — you'd have to substitute what y is equivalent to from the parabola into the equation for the circle. It gets a bit messy. But, because I see x^2 and $-x^2$, $6x$ and $-6x$ in the two equations, I'm going to take advantage of the situation and use elimination.

First, set the equation of the parabola equal to 0, rearrange the terms in both, and add the two equations together:

$$\begin{aligned} 0 &= -x^2 + 6x \qquad\quad - y + 8 \\ 0 &= \ \ x^2 - 6x + y^2 - 8y \\ \hline 0 &= \qquad\qquad\quad y^2 - 9y + 8 \end{aligned}$$

Now factor the quadratic for y:

$$\begin{aligned} 0 &= y^2 - 9y + 8 \\ &= (y-1)(y-8) \\ y &= 1 \text{ or } y = 8 \end{aligned}$$

Substituting 1 in for y in the equation of the parabola and solving the resulting quadratic in x, you get:

$$1 = -x^2 + 6x + 8$$
$$x^2 - 6x - 7 = 0$$
$$(x - 7)(x + 1) = 0$$
$$x = 7 \text{ or } x = -1$$

So, when $y = 1$, x is either 7 or –1. This gives you two solutions: (7, 1) and (–1, 1). You go through a similar process with $y = 8$ and get that $x = 0$ or $x = 6$. So the final two points of intersection are at (0, 8) and (6, 8).

A circle and a parabola can also intersect at three points, two points, one point, or no points.

You use the same methods to solve systems of equations that end up with fewer than four intersections. The algebra leads you to the solutions — but beware the false promises. You have to watch out for extraneous solutions by checking your answers.

If substituting one equation into another, take a look at the resulting equation. The highest power of the equation tells you what to expect as far as the number of common solutions. When the power is 3 or 4, you can have as many as three or four solutions, respectively. When the power is 2, you can have up to two common solutions. A power of 1 indicates only one possible solution. If you end up with an equation that has no solutions, you know the system has no points of intersection — the graphs just pass by like ships in the night.

Sifting through the possibilities for solutions

In the "Intersecting Parabolas and Lines" section, earlier in this chapter, the examples I provide use substitution where the x's replace the y variable. Most of the time, this is the method of choice, but I suggest you remain flexible and open for other opportunities. The next example is just such an opportunity — taking advantage of a situation where it makes more sense to replace the x term with the y term.

Find the common solutions of the parabola $y = x^2$ and the circle $x^2 + (y - 1)^2 = 9$.

Take advantage of the simplicity of the equation $y = x^2$ by replacing the x^2 in the circle equation with y. That sets you up with an equation of y's to solve:

$$\begin{aligned} x^2 + (y-1)^2 &= 9 \text{ and } y = x^2 \\ y + (y-1)^2 &= 9 \\ y + y^2 - 2y + 1 &= 9 \\ y^2 - y - 8 &= 0 \end{aligned}$$

This quadratic equation doesn't factor, so you have to use the quadratic formula to solve for y:

$$y = \frac{1 \pm \sqrt{1 - 4(1)(-8)}}{2(1)} = \frac{1 \pm \sqrt{33}}{2}$$

You find two different values for y, according to this solution. When you use the positive part of the $\pm$, you find that y is close to 3.37. When you use the negative part, you find that y is about –2.37. Something doesn't seem right. What is it that's bothering you? It has to be the negative value for y. The common solutions of a system should work in both equations, and $y = -2.37$ doesn't work in $y = x^2$, because when you square x, you don't get a negative number. So, only the positive part of the solution, where $y \approx 3.37$, works.

Substitute $\frac{1+\sqrt{33}}{2}$ into the equation $y = x^2$ to get x:

$$\begin{aligned} \frac{1+\sqrt{33}}{2} &= x^2 \\ \pm\sqrt{\frac{1+\sqrt{33}}{2}} &= x \end{aligned}$$

The value of x comes out to about ± 1.84. The graph in Figure 11-2 shows you the parabola, the circle, and the points of intersection at about (1.84, 3.37) and about (–1.84, 3.37).

When $y = -2.37$, you get points that lie on the circle, but these points don't fall on the parabola. The algebra shows that, and the picture agrees.

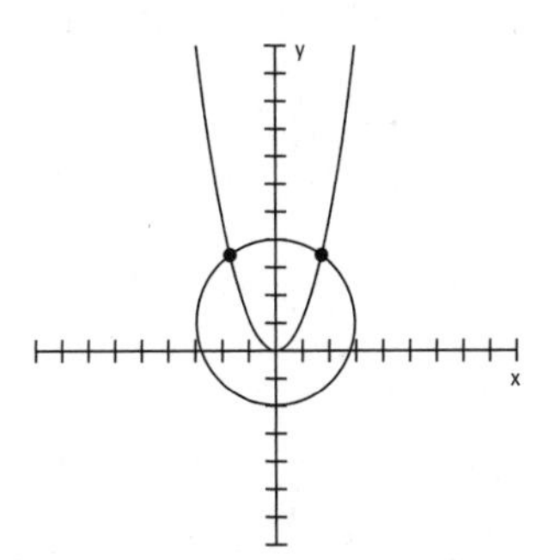

Figure 11-2: This system has only two points of intersection.

When substituting into one of the original equations to solve for the other variable, always substitute into the *simpler* equation — the one with smaller exponents. This helps you catch any extraneous solutions.

Chapter 12

Taking the Complexity Out of Complex Numbers

In This Chapter

- Defining imaginary and complex numbers
- Writing complex solutions for quadratic equations
- Determining complex solutions for polynomials

Mathematicians define *real numbers* as all the whole numbers, negative and positive numbers, fractions and decimals, radicals — anything you can think of to use in counting, graphing, and comparing amounts. Mathematicians introduced imaginary numbers when they couldn't finish some problems without them. For example, when solving for roots of quadratic equations such as $x^2 + x + 4 = 0$, you quickly discover that you can find no real answers. Using the quadratic formula, the solutions come out to be

$$x = \frac{-1 \pm \sqrt{1^2 - 4(1)(4)}}{2(1)} = \frac{-1 \pm \sqrt{-15}}{2}$$

The equation has no real solution. So, instead of staying stuck there, mathematicians came up with something innovative. They made up a number and named it i.

The square root of –1 can be replaced with the imaginary number i: $\sqrt{-1} = i$. Furthermore, $i^2 = -1$.

In this chapter, you find out how to create, work with, and analyze imaginary numbers and the complex expressions they appear in. Just remember to use your imagination!

Simplifying Powers of i

The powers of i (representing powers of imaginary numbers) follow the same mathematical rules as the powers of real numbers. The powers of i, however, have some neat features that set them apart from other numbers.

You can write all the powers of i as one of four different numbers: i, $-i$, 1, and -1; all it takes is some simplifying of products, using the properties of exponents, to rewrite the powers of i:

- $\boldsymbol{i = i}$**:** Just plain old i.
- $\boldsymbol{i^2 = -1}$**:** From the definition of an imaginary number (see the introduction to this chapter).
- $\boldsymbol{i^3 = -i}$**:** Use the rule for exponents ($i^3 = i^2 \cdot i$) and then replace i^2 with -1. So, $i^3 = (-1) \cdot i = -i$.
- $\boldsymbol{i^4 = 1}$**:** Because $i^4 = i^2 \cdot i^2 = (-1)(-1) = 1$.
- $\boldsymbol{i^5 = i}$**:** Because $i^5 = i^4 \cdot i = (1)(i) = i$.
- $\boldsymbol{i^6 = -1}$**:** Because $i^6 = i^4 \cdot i^2 = (1)(-1) = -1$.
- $\boldsymbol{i^7 = -i}$**:** Because $i^7 = i^4 \cdot i^2 \cdot i = (1)(-1)(i) = -i$.
- $\boldsymbol{i^8 = 1}$**:** Because $i^8 = i^4 \cdot i^4 = (1)(1) = 1$.

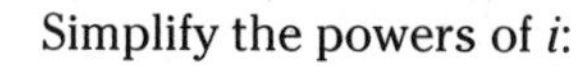

Simplify the powers of i:

- $\boldsymbol{i^{41} = i}$**:** Because $i^{41} = i^{40} \cdot i = (i^4)^{10}(i) = (1)^{10} \cdot i = 1 \cdot i = i$.
- $\boldsymbol{i^{935} = -i}$**:** Because $i^{935} = i^{932} \cdot i^3 = (i^4)^{233}(i^3) = (1)^{233}(-i) = 1(-i) = -i$.

Every power of i where the exponent is a multiple of 4 is equal to 1. If the exponent is one value greater than a multiple of 4, the power of i is equal to i. An exponent that's two more than a multiple of 4 results in -1; and three more than a multiple of 4 as a power of i results in $-i$. So, all you need do to change the powers of i is figure out where the exponent is in relation to some multiple of four.

Getting More Complex with Complex Numbers

The imaginary number i is a part of the numbers called *complex numbers*, which arose after mathematicians established imaginary numbers. The standard form of complex numbers is $a + bi$, where a and b are real numbers, and i^2 is –1. The fact that i^2 is equal to –1 and i is equal to $\sqrt{-1}$ is the foundation of the complex numbers.

Some examples of complex numbers include $3 + 2i$, $-6 + 4.45i$, and $7i$. In the last number, $7i$, the value of a is 0.

Performing complex operations

You can add, subtract, multiply, and divide complex numbers — in a very careful manner. The rules used to perform operations on complex numbers look very much like the rules used for any algebraic expression, with two big exceptions:

- You simplify the powers of i.
- You don't really divide complex numbers — you change the division problem to a multiplication problem.

Making addition and subtraction complex

When you add or subtract two complex numbers $a + bi$ and $c + di$ together, you get the sum (difference) of the real parts and the sum (difference) of the imaginary parts:

$$(a + bi) + (c + di) = (a + c) + (b + d)i$$

$$(a + bi) - (c + di) = (a - c) + (b - d)i$$

Add $(-4 + 5i)$ and $(3 + 2i)$; then subtract $(3 + 2i)$ from $(-4 + 5i)$.

$$(-4 + 5i) + (3 + 2i) = (-4 + 3) + (5 + 2)i = -1 + 7i$$

$$(-4 + 5i) - (3 + 2i) = (-4 - 3) + (5 - 2)i = -7 + 3i$$

Creating complex products

To multiply complex numbers, you have to treat the numbers like binomials and distribute both the terms of one complex number over the other:

$$(a + bi)(c + di) = (ac - bd) + (ad + bc)i$$

Find the product of $(-4 + 5i)$ and $(3 + 2i)$.

$$(-4 + 5i)(3 + 2i) = -12 - 8i + 15i + 10i^2$$

You simplify the last term by replacing the i^2 with -1 to give you -10. Then combine -10 with the first term. Your result is $-22 + 7i$, a complex number.

Performing complex division by multiplying by the conjugate

The complex thing about dividing complex numbers is that you don't really divide. Instead of dividing, you do a multiplication problem — one that has the same answer as the division problem.

Describing the conjugate of a complex number

A complex number and its *conjugate* are $a + bi$ and $a - bi$. The real part, the a, stays the same; the sign between the real and imaginary part changes. For example, the conjugate of $-3 + 2i$ is $-3 - 2i$, and the conjugate of $5 -3i$ is $5 + 3i$.

The product of an imaginary number and its conjugate is a real number (no imaginary part) and takes the following form:

$$(a + bi)(a - bi) = a^2 + b^2$$

Dividing complex numbers

When a problem calls for you to divide one complex number by another, you write the problem as a fraction and then multiply by a fraction that has the conjugate of the denominator in both numerator and denominator.

Divide $(-4 + 5i)$ by $(3 + 2i)$.

Write the problem as a fraction. Then multiply the problem's fraction by a second fraction that has the conjugate of $3 + 2i$ in both numerator and denominator.

$$\begin{aligned}\frac{-4+5i}{3+2i}\cdot\frac{3-2i}{3-2i} &= \frac{-12+8i+15i-10i^2}{3^2+2^2}\\ &= \frac{-12+(8+15)i-10(-1)}{9+4}\\ &= \frac{-12+10+(8+15)i}{13}\\ &= \frac{-2+23i}{13}\\ &= -\frac{2}{13}+\frac{23}{13}i\end{aligned}$$

Simplifying reluctant radicals

Until mathematicians defined imaginary numbers, many problems had no answers because the answers involved square roots of negative numbers. After the definition of an imaginary number, $i^2 = -1$, came into being, the problems involving square roots of negative numbers were solved.

To simplify the square root of a negative number, you write the square root as the product of square roots and simplify: $\sqrt{-a} = \sqrt{-1}\sqrt{a} = i\sqrt{a}$.

Simplify $\sqrt{-24}$.

First, split up the radical into the square root of –1 and the square root of the rest of the number, and then simplify by factoring out perfect squares:

$$\sqrt{-24} = \sqrt{-1}\sqrt{24} = \sqrt{-1}\sqrt{4}\sqrt{6} = i\cdot 2\sqrt{6}$$

By convention, you write the previous solution as $2i\sqrt{6}$.

Unraveling Complex Solutions in Quadratic Equations

You can always solve quadratic equations with the quadratic formula. It may be easier to solve quadratic equations by factoring, but when you can't factor, the formula comes in handy. Until mathematicians began recognizing imaginary numbers, however, they couldn't complete many results of the quadratic formula. Whenever a negative value appeared under a radical, the equation stumped the mathematicians.

The modern world of imaginary numbers to the rescue! Now quadratics with complex answers have results to show.

Solve the quadratic equation $2x^2 + x + 8 = 0$.

Using the quadratic formula, you let $a = 2$, $b = 1$, and $c = 8$:

$$\begin{aligned} x &= \frac{-1 \pm \sqrt{1^2 - 4(2)(8)}}{2(2)} \\ &= \frac{-1 \pm \sqrt{1 - 64}}{4} \\ &= \frac{-1 \pm \sqrt{-63}}{4} \\ &= \frac{-1 \pm \sqrt{-1}\sqrt{9}\sqrt{7}}{4} \\ &= \frac{1 \pm 3i\sqrt{7}}{4} \end{aligned}$$

Investigating Polynomials with Complex Roots

Polynomials are functions whose graphs are nice, smooth curves that may or may not cross the x-axis. If the degree (or highest power) of a polynomial is an odd number, its graph must cross the x-axis, and it must have a real root or solution. When solving equations formed by setting polynomials equal to 0, you plan ahead as to how many solutions you can expect to find. The highest power tells you the maximum number of solutions you can find. If the solutions are real, then the curve either crosses the x-axis or touches it. If any solutions are complex, then the number of crossings or touches is decreased by the number of complex roots.

Classifying conjugate pairs

A polynomial of degree (or power) n can have as many as n real zeros (also known as solutions, roots, or x-intercepts). If the polynomial doesn't have n real zeros, it has $n - 2$ zeros, $n - 4$ zeros, or some number of zeros decreased two at a time. The reason that the number of zeros decreases by two is that complex zeros always come in *conjugate pairs* — a complex number and its conjugate.

Complex zeros, or solutions of polynomials, come in *conjugate pairs* — $a + bi$ and $a - bi$. If one of the pair is a solution, then so is the other.

The equation $0 = x^5 - x^4 + 14x^3 - 16x^2 - 32x$, for example, has three real roots and two complex roots, which you know because you apply the rational root theorem and Descartes' rule of signs (see Chapter 7) and ferret out those real and complex solutions. The equation factors into $0 = x(x - 2)(x + 1)(x^2 + 16)$. The three real zeros are 0, 2, and –1. The two complex zeros are $4i$ and $-4i$. You say that the two complex zeros are a *conjugate pair,* and you get the roots by solving the equation $x^2 + 16 = 0$.

Making use of complex zeros

The polynomial function $y = x^4 + 7x^3 + 9x^2 - 28x - 52$ has two real roots and two complex roots. According to Descartes' rule of signs, the function could've contained as many as four real roots (suggested by the rational root theorem). You can determine the number of complex roots in two different ways: by factoring the polynomial or by looking at the graph of the function.

The polynomial function factors into $y = (x - 2)(x + 2)(x^2 + 7x + 13)$. The first two factors give you real roots, or x-intercepts. When you set $x - 2$ equal to 0, you get the intercept (2, 0). When you set $x + 2$ equal to 0, you get the intercept (–2, 0). Setting the last factor, $x^2 + 7x + 13$, equal to 0 doesn't give you a real root.

But you can also tell that the polynomial function has complex roots by looking at its graph. You can't tell what the roots are, but you *can* see that the graph has some. If you need the values of the roots, you can resort to using algebra to solve for them. Figure 12-1 shows the graph of the example function, $y = x^4 + 7x^3 + 9x^2 - 28x - 52$. You can see the two x-intercepts, which represent the two real zeros. You also see the graph flattening on the left.

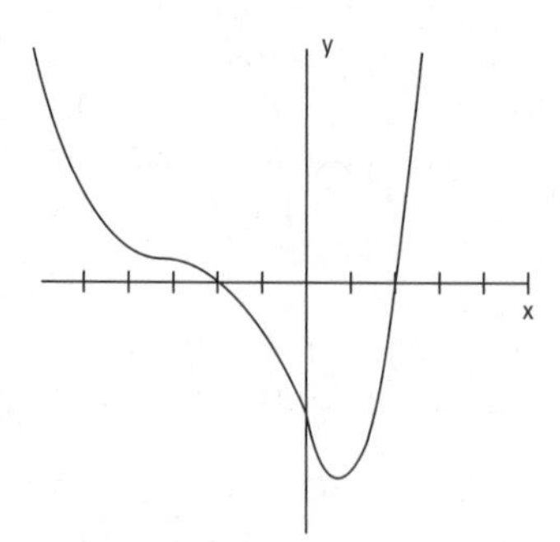

Figure 12-1: A flattening curve indicates a complex root.

Figure 12-2 can tell you plenty about the number of real zeros and complex zeros the graph of the polynomial has . . . before you ever see the equation it represents.

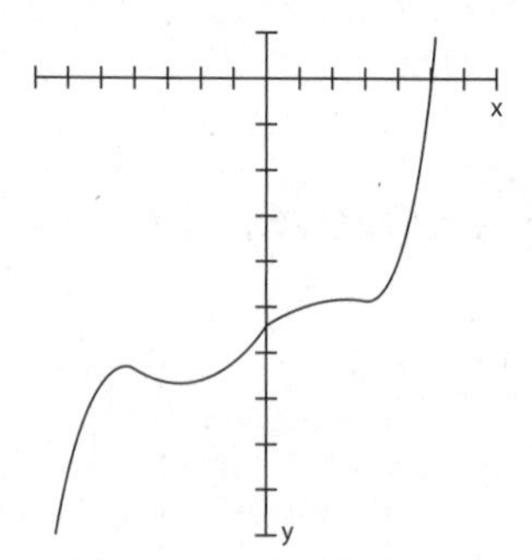

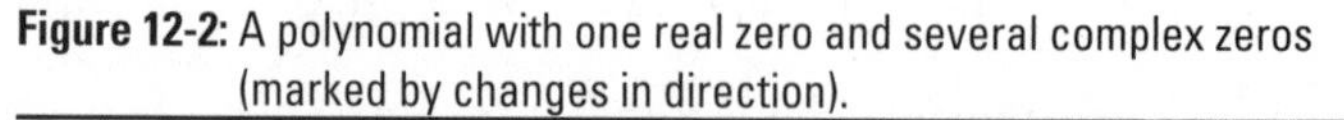

Figure 12-2: A polynomial with one real zero and several complex zeros (marked by changes in direction).

The polynomial in Figure 12-2 appears to have one real zero and several complex zeros. Do you see how it changes direction all over the place under the x-axis? These changes indicate the presence of complex zeros. The graph represents the polynomial function $y = 12x^5 + 15x^4 - 320x^3 - 120x^2 + 2880x - 18{,}275$. The function has four complex zeros — two complex (conjugate) pairs — and one real zero (when $x = 5$).

Chapter 13

Ten (Or So) Special Formulas

In This Chapter

- Counting the number of items or groups or arrangements
- Adding up large lists of numbers
- Figuring interest earned or interest paid

A formula is actually an equation that expresses some relationship that always holds true. In this chapter, you find ten (or so) formulas that are found frequently in algebra and some mathematical studies that use a lot of algebra.

Using Multiplication to Add

The *multiplication property of counting* states that if you choose one item from the first set of choices, one item from the second set of choices, one item from the third set of choices, and so on, all you need do to find the total number of arrangements you might create is to multiply how many items are in each set.

So, if you have ten shirts, six pairs of slacks, eight pairs of socks, and three pairs of shoes, you can determine the total number of different outfits possible. Just multiply $10 \cdot 6 \cdot 8 \cdot 3 = 1,440$. You won't have to repeat an outfit for several years!

Factoring in Factorial

The *factorial* operation says that you take a whole number and multiply it times every natural number smaller than that whole number: $n! = n \cdot (n-1) \cdot (n-2) \ldots 3 \cdot 2 \cdot 1$. Also, by special designation, $0! = 1$.

To find 6!, you multiply $6 \cdot 5 \cdot 4 \cdot 3 \cdot 2 \cdot 1 = 720$.

Picking Out Permutations

A *permutation* is a way of counting how many different arrangements are possible if you choose *r* items out of a possible *n* items and need them in a particular order:

$${}_nP_r = \frac{n!}{(n-r)!}$$

So, if you have five finalists in a race and want to determine how many different ways you can have first and second place happen, you find the number of permutations possible with 5 choose 2:

$${}_5P_2 = \frac{5!}{(5-2)!} = \frac{5!}{3!} = \frac{5 \cdot 4 \cdot \cancel{3} \cdot \cancel{2} \cdot \cancel{1}}{\cancel{3} \cdot \cancel{2} \cdot \cancel{1}} = 5 \cdot 4 = 20$$

The computation gives you the *number* of arrangements. Now you have to list them: Andy and Bob, Andy and Chuck, and so on.

Collecting Combinations

A *combination* is a way of counting how many different arrangements are possible if you choose *r* items out of a possible *n* items where the order they're in doesn't matter:

$${}_nC_r = \frac{n!}{r!(n-r)!}$$

As you may have noticed, the only difference between permutations and combinations is that the denominator in the formula for combinations has the additional factor — making the denominator a larger number (if r isn't 0 or 1).

Adding n Integers

When you want to add $1 + 2 + 3 + 4 + \ldots + n$, use the following formula:

$$\sum_{i=1}^{n} i = \frac{n(n+1)}{2}$$

So, if the bottom row of your stack has 20 blocks, the next row up has 19 blocks, and so on, the total number of blocks in your stack is

$$\sum_{i=1}^{20} i = \frac{20(20+1)}{2} = 210 \text{ blocks}$$

Adding n Squared Integers

When you want to add $1^2 + 2^2 + 3^2 + 4^2 + \ldots + n^2$, use the following formula:

$$\sum_{i=1}^{n} i^2 = \frac{n(n+1)(2n+1)}{6}$$

To find the sum of the first ten squares, $1^2 + 2^2 + 3^2 + 4^2 + \ldots + 10^2$:

$$\sum_{i=1}^{10} i^2 = \frac{10(10+1)[2(10)+1]}{6} = 385$$

Adding Odd Numbers

When you want to add $1 + 3 + 5 + 7 + \ldots + (2n - 1)$, use the following formula:

$$\sum_{i=1}^{n} (2i-1) = n^2$$

Computing the sum of the first ten odd numbers, 1 + 3 + 5 + 7 + . . . + 19:

$$\sum_{i=1}^{10}(2i-1)=10^2=100$$

Going for the Geometric

A geometric sequence is formed by multiplying by the same number repeatedly. For example, multiplying by the number three, over and over again: 1, 3, 9, 27, 81, 243, ... To add up all the terms in a geometric sequence, you use the formula:

$\sum_{i=1}^{n} ar^{i-1}=\frac{a(1-r^n)}{1-r}$, where a is the first term in the sequence and r is the ratio or repeating multiplier.

So the sum of 1, 3, 9, 27, 81, and 243 is

$$\begin{aligned}\sum_{i=1}^{6} 1\cdot 3^{i-1} &= \frac{1(1-3^6)}{1-3}\\ &= \frac{-728}{-2}\\ &= 364\end{aligned}$$

And, to add to the excitement, here's the formula for finding the sum of an *infinite* geometric sequence — all the terms forever and ever:

$$\sum_{i=1}^{\infty} ar^{i-1}=\frac{a}{1-r}$$

The formula only works if the multiplier, r, is between –1 and 1.

Calculating Compound Interest

You deposit $10,000 in an account that earns 2 percent interest, compounded quarterly. How much will there be in the account at the end of 20 years? Use the following formula:

$A = P\left(1+\frac{r}{n}\right)^{nt}$, where A is the total amount accumulated, P is the principal or beginning amount, r is the interest rate written as a decimal, n is the number of times compounded per year, and t is the number of years

So, to answer the opening question, you'd have

$$A = 10{,}000\left(1+\frac{0.02}{4}\right)^{4(20)} \approx \$14{,}903.39$$

Index

• Symbols •

• A •

• B •

• C •

• D •

• E •

• F •

• M •

• N •

• O •

• P •